COMPREHENSIVE REVIEW OF THE
ELECTRONICS

(ANALOG, DIGITAL, MICROPROCESSOR)

DR. MGA SIDDIQUI

PH.D. (PHYSICS). CIC, PGDCA, RHCT, CNE, ADCHN

BLUEROSE PUBLISHERS

India | U.K.

For permissions requests or inquiries regarding this publication,
please contact:

BLUEROSE PUBLISHERS
www.BlueRoseONE.com
info@bluerosepublishers.com
+91 8882 898 898
+4407342408967

ISBN: 978-93-5989-553-6

Cover design: Tahira
Typesetting: Tanya Raj Upadhyay

First Edition: February 2024

ELECTRONICS

(Analog, Digital, Microprocessor)

For the students of
Class-XI (Level-3) and Class-XII (Level-4),
Diploma Engineering, B.Sc., B. Tech., BCA,
and 14^+ Skill/Vocational/ PMKVY Courses
in Electronics & ESDM Sector

Under
National Education Policy - 2020 (NEP-2020)
Govt. of India

AUTHOR
DR. MGA SIDDIQUI
Ph.D.(Physics), CIC, PGDCA, RHCT, CNE, ADCHN
Delhi, India

---

www.iict-india.org   www.gram-infotech.com

Preface

Technologies like digital humans, satellite communications, tiny ambient IoT, secure computation and autonomic robots will change the human life. Besides, disruptive technologies like Artificial Intelligence (AI), Machine learning (ML), Robotic Process Automation (RPA), Data Science, DevOp, quantum computing ,Block chain, Edge Computing.

This book will provide a practical approach with basics of electronics to the students at **Senior Secondary level in class-XI (Level-3) and class-XII (Level-4) in their Vocational Courses.** It is as per the motive of National Education Policy-2020 (NEP-2020), the Government of India for skilling India, and boosts the youth towards Make in India, the dream of our nation

This book is intended for students taking an introductory electronics course. **Current syllabi of basic electronics included in Physics (Honours), Diploma, B.Tech., BCA,** curriculum of different universities and those offered in various engineering and technical institutions.

It will also help those who are going **for 14^+ Skill/Vocational/PMKVY** courses in the electronics sector and have no science background. The ESDM (Electronics System Design and Manufacturing) sector, focusing on students and unemployed youth at 9–10th standard onwards, ITI, diploma, non-engineering graduates, etc., to increase their employability to work in 'Manufacturing' and 'Service Support' functions for skill development in ESDM sector.

Fundamental ideas are the foundation of all advanced technology. It is based on speed, accuracy and miniaturization of a semiconductor chip. The invention of Electron was breakthrough towards the modernized shape of Electrical, Analog Electronics, Digital Electronics and Microprocessor. From Electron to Electronics, from Diode to Transistor, from Transistor to Logic Gates from Logic Gates to Chips (Integrated Circuits), from group of ICs to Microprocessor, From microprocessor to Super computer, From Super computer to Quantum computing.

This book has been composed in a way that students can get the concept of basics in sequence of three sections as **PART- I: Analog Electronics, PART-II: Digital Electronics and PART-III: Microprocessor.**

PART-I: Analog Electronics: This part encompasses the evolution of electronics, atoms and element, electric current, an understanding of ohm's law , electric motive force, composition of transformers etc. where fundamentals of resistors, potentiometers , capacitors ,inductors, voltage source, battery, , transistors, etc., have been discussed. Furthermore, it emphasizes the fundamentals of semiconductors, providing an understanding of metals, semiconductors, insulators, rectifiers, PN Junction diodes, and so forth. The content provides information about various configurations of Junction Transistor, JFET, UJT, Transistor amplifier and its applications, SCR, VRACTOR DIODE and ZENER DIODE.

PART-II: Digital Electronics : The part deals with Properties Of Logic Gates and Boolean Algebra, Binary number system, Flip-Flop, Counters and Registers, Multiplexer, Demultiplexer , Encoder, Decoder , Logic Minimization & Karnaugh Maps, Converters , Memory Terminology Semiconductors;

PART-III: Microprocessor : The section is intended for introductory microcontroller courses at the under graduate level in technology and engineering. It covers Introduction to Microprocessor, Microprocessor & Its Architecture, Programming Model of 8085 Microprocessor, 8085/8085A μP Assembler Language Programming, Interfacing with Microprocessor. it is very useful to every electronics and computer science students..

This book's language is straightforward and basic enough for students to understand. In the area of electronics and communication, examples and condensed concepts make learning easy and comfortable for the student. The student can use this book as a tool for easy learning.

This book is authored by 25+ years of experience as an entrepreneur/Trainer, the CEO/FOUNDER/PRESIDENT of the Indian Institute of Computer Technology (IICT), Graminfotech Pvt. Ltd., and the Society for Computer Educational Research, having a doctorate degree in physics with various international publications and workshops and bearing some professional certifications of IT services like training, development, and consultancy. Author of 4 Books in Electronics, Computer Peripherals—Part I, Computer Peripherals—Part II, and Computer Networking—Do it.

Every care has been taken to check mistakes and misprints, yet it is very difficult to claim perfection. Any undetected and unintentional errors, omissions, suggestions etc. from students and colleagues for improvement brought to my notice in good spirit are always most welcome.

Although I enjoy using subjective approaches, it is not my main goal. It is my aim that this book will inspire the student within each of them to learn about and enjoy electronics.

AUTHOR
Dr. MGA SIDDIQUI
Ph.D. (Physics), CIC, PGDCA, RHCT, CNE, ADCHN
Delhi, India
https://gram-infotech.com/founder.php#
Email: drghufranphysics@gmail.com
Whatsapp: +19- 9897278615

Acknowledgments

First and foremost, I would want to express my gratitude for the inspiration and support I received from the faculty and staff at the Indian Institute of Computer Technology (www.iict-india.org) as well as from my friends and colleagues. It was the confined space that prevented me from calling those.

For their assistance at various stages of the book's development, I am grateful to my closest relatives. In writing this book over the course of eleven months, I must acknowledge the assistance and support I had from my wife (home minister). Although I put a lot of work into this book, they all contributed time.

In the more than twenty years I've worked as a entrepreneur, my responsibilities have included curriculum development, teaching, training, designing electronics equipment for skill courses in analog, digital electronics, microprocessors, computer peripherals, and networking, all with the mission "FROM DARKNESS TO BRIGHTNESS" (FROM UNEMPLOYMENT TO EMPLOYMENT). Since 1997, I've interacted with numerous groups of students and business professionals, and I've come to appreciate the value of such interaction. It has also been quite good to interact with the current PMKVY Training programs for students run at my Institute (IICT).

HUMANITY THE ONLY RELIGION ???

The only groups on earth are male and female, and humanity as a whole has only one religion. All other religions and groups are fictitious, made up and controlled by religious authorities for their own personal benefit.

At this point, Life is nothing but to serve the humanity without any discrimination. We must always establish a justice in our family and then whole of the universe as per our ability." We cannot serve the humanity if we engage our self in any human made religion, which creates discrimination.

If any human creating discrimination among human on the basis of religion then it is self doctrine by human for their own personal benefits. You must overcome from all types of religious rituals to establish a peace in the universe. We should learn from nature, it gives every one without any discrimination like sun light and shower of rain for every creature on the face of earth.

We should give to everyone like nature, if we have a power; give the justice to the victim without any delay. If we have power of money then help those who are waiting, start from your dears and nears.

Thank you once again to everyone who helped me with the difficult task of writing a book, whether it was with their financial, moral, academic, or emotional assistance.

AUTHOR

DR. MGA SIDDIQUI

Delhi, India

Ph.D.(Physics), CIC, PGDCA, RHCT, CNE, ADCHN

https://gram-infotech.com/founder.php#

CONTENTS
PART I -ANALOG ELECTRONICS

CHAPTER 4

Electric Current   **29**

CHAPTER 5

Magnetism   **49**

CHAPTER-6

Semiconductors   58

CHAPTER-7

PN- Junction Diode   65

CHAPTER-8

Transistor   **74**

CHAPTER-9

Specialized Devices   **82**

PART II -DIGITAL ELECTRONICS

CHAPTER-10
Introduction

CHAPTER-11
Logic Gates and Boolean Algebra   94

CHAPTER-12
Binary number system   100

PART III -MICROPROCESSOR

CHAPTER-23

Interfacing with Microprocessor   196

PART-I

Analog Electronics

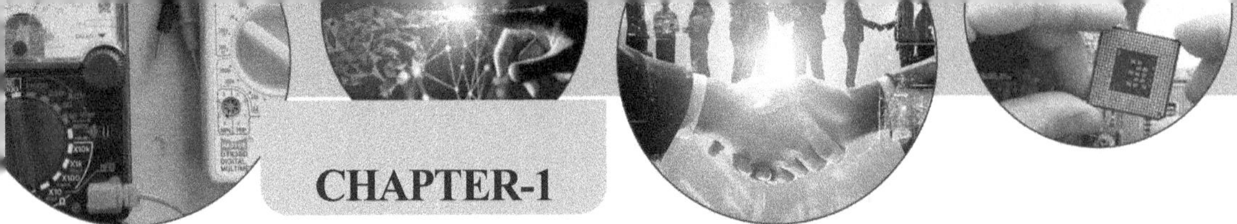

CHAPTER-1

Introduction to Electrical Charge, Current and Voltage

1.1 Introduction: The basis of modern electronics has evolved from the work and discoveries of a large number of scientists during the past three centuries. In the latter part of the seventeenth century, Sir Isaac Newton formulated the laws which describe the motion of material bodies. During the eighteenth centuries, the laws of electricity and magnetism were discovered.

They were synthesized and summarized in the latter part of the nineteenth century by James Clerk Maxwell in his electromagnetic theory. At the beginning of the present century, the concepts of electrical charge, current and voltage were firmly established. Classical physics, consisting of Newton mechanics and Maxwell electromagnetism, was considered at the beginning of the present century to provide a complete understanding of our physical environment However, new discoveries, soon began to point out the inadequacy of classical physics. In 1897 the electron was identified. In 1905. Albert Einstein introduced new concepts of space and time in his theory of relativity.

Soon after, Ernest Rutherford and Neils Bohr advanced the first approximations of the true structure of the basic structure of matter followed. Protons were identified as the carriers of positive electrical charge. In 1932 the neutron was discovered. The vacuum tube had been introduced in 1905, the first two and one half decades the spectrum of modern electronics has been completed by the discovery of microwaves and the introduction of the fascinating maser and laser devices. All of these discoveries and developments were vital in the growth of electronics.

1.2 THE HISTORICAL BACKGROUND OF ELECTRONICS

1.2.1 Sir Isaac Newton (1642-1727): He was the first to formulate the laws of motion of material bodies. Further, he formalized the concept of mass, force, momentum, acceleration and velocity and he introduced the theory of gravitation. In terms of these theories he was able to adequately describe the motion of the planets around the sun. The same laws describe much of the behavior of electrons in electronic devices. A unit of force, the Newton is named in this honour.

1.2.2 Charles Augustin de Coulomb (1736–1806): He was born in France nine years after Newton's demise. He served as an engineer in the French military for nine years until his health was impaired in the west Indies. He retired to a small estate in Blois where he conducted his studies. In 1785, after he had invented a sensitive balance for measuring small forces, Coulomb experimentally verified the laws of electrical repulsion and attraction and observed the inverse square nature of the forces between electrical charge. Because of his pioneering work, a unit of electrical charge is called the Coulomb. During the same period of time, in Scotland.

1.2.3 James Watt (1736-1819): He was conducting his experiments with steam engines and mechanical power. Though not directly a contributor to the development of our understanding of electricity or magnetism, his inventive contribution in the development of practical steam engines resulted in naming a unit of power (mechanical or

electrical) the Watt. He and an associate, Matthew Boulton, coined the term horsepower as a unit of mechanical power.

1.2.4   Alessandro Volta (1745-1827): An Italian physicist, was another who made significant early contribution to experimental electricity. He developed the voltaic pile, the fore runner of the modern battery, at the close of the eighteenth century. It provided subsequent experimenters with a convenient source of voltage and current. In 1779 Volta became professor of physics at the University of Pavia, where he remained of his work, and the term volt, applied to electrical voltage is named in his honour.

1.2.5   Hans Christian Oersted (1777-1851): One of the most important discoveries in electricity and magnetism. About he observed that the needle of magnetism compass was deflected when in the vicinity of a current carrying wire. This was the first suggestion of a relationship between electricity(the current) and magnetism (the needle). Oersted was born in Denmark and earned the degree of Doctor of Philosophy at the University of Copenhagen.

1.2.6 Andre marie Ampere (1775-1836): A magnetism field quantity, the Oersted is named in his honour. A French physical, made significant advances in the mathematical theory of electricity and magnetism and in identifying one with the other in electromagnetism. A short time after learning of Oersted's important discovery. Ampere demonstrated the magnetic interaction of two current carrying conducted while he was professor of physics at the Ecole Polytechnic, a position he held from 1809 until his death. In his honor, the ampere, a measure of electrical current, was created.

1.2.7 Karl Friend rich Gauss (1777-1855): A German, who is often reffered to as the greatest mathematician of all time. Gauss's principle work was in mathematical physics, the application of mathematics to electricity and magnetism was among his most important application of mathematics to electricity and magnetism was among his most important contributions.

1.2.8 Goerge Simon (1787-1854): Ohm's Law and the unit of electrical resistance are both named in honour of Goerge Simon Ohm a German scientist. He received his education at the Erlangen University. In 1817 he became a teacher at the gymnasium in Cologne and in 1826 in Berlin. He taught in Nuremberg and later in Munich as professor of physics. His researches were chiefly concerned with electrical currents and voltage and he formulated Ohm's Law relating current, voltage and resistance.

1.2.9 Michael Faraday (1791-1867) may justly be called one of the greatest experimentalist. Born in England, the son of the blacksmith, he spent his early years as an apprentice bookbinder. Faraday was largely self educated. He attended a series of lectures on chemistry presented by Sir Humphery Davy, lectures to Davy, Faraday was appointed an assistant in the laboratory of the Royal institution. He was named laboratory director in 1825.

He made many contributions in chemistry but his most important discoveries were in electricity and magnetism. He demonstrated the generation of electrical currents and voltages by changing magnetic field, thus contributing to the inseparable union of electricity and magnetism as well as providing the key to such fundamental devices as generators, transformers and inductors.

This remarkable experimental scientist has a unit of charge, the Faraday and a unit of capacitance, the Farad named for him.

1.2.10 Joseph Henry (1797- 1878): The first great American scientist was Joseph Henry whose work in electricity and magnetism closely paralleled and sometimes exceeded that of Faraday in England. Henry directed his early studies toward the medical profession, but an appointment in 1825 to survey a route from the Hudson River to lake Erie changed his interest to engineering and his experimental genius to electricity and magnetism He experimented with large electromagnets, invented and demonstrated the first telegraph and devised and constructed the fore runner of electric motors. Henry was the name given to the inductance unit.

1.2.11 Wilhelm Eduard (1804-1891): He held a professorships Princeton, was one of the original members and president of National Academy of Science, was secretary of the Smithsonian Institute and brought recognition to the United States Weather Services. A German physicist Wilhelm Eduard Weber also made various contribution to the development of the theory of electricity and magnetism.

His work was significant in demonstrating that the units of electrical and magnetic quantities may be expressed in terms of length, mass and time. Weber's further work with units encouraged Maxwell to embark upon the mathematical study which culminated in the prediction of the existence of electromagnetic waves, a prediction verified later by Heinrich Hertz. Weber was professor of physics at Gottingen and Leipzig and was an associate of Gauss. Together in 1833 they constructed a telegraph and conducted investigations of terrestrial magnetism. A unit of magnetic flux, the Weber is named in his honour.

1.2.12 James Prescott Joule (1818-1889): Energy is a particularly important physical quantity in electronics. An English physicist is largely responsible for our present understanding of energy conversion and conservation. In 1843 he announced his determination of the amount of mechanical work required to generate a unit of heat, thus demonstrating the conversion of mechanical energy into thermal (heat) energy. His many contributions to science, especially in the fields of electricity and thermodynamics win him due recognition.

A unit of energy, the Joule is named in his honour. Thus, with the work of Maxwell knowledge of electricity and magnetism had a sound basis and all the observed phenomena were well explained. However, modem electronics was as yet nonexistent and was not to have its real beginning until the physical embodiment of electrical charge and its behaviour were understood.

Prior to 1891 the laws of electrolysis had suggested the atomicity of electricity and had related chemical reactions and electricity. In that year G. Johnstone Stoney proposed the name electron for the atomic unit of charge. Earlier in 1879, Sir William Croocks had

observed and described cathode rays in electrical discharges in low pressure gases. Jean Baptiste Perrin in 1895 passed these cathode rays into an insulated chamber attached to a device for measuring charge and proved that they carried a negative charge.

1.2.13 J.J. Thomson (1856-1940): The greatest single step toward an understanding of the cathode rays and their identification as electrons was made in the famous experiments of J.J. Thomson culminating in 1897 which is thus attributed with the 'discovery' of the electrons, although neither charge nor mass was known individually.

1.2.14 R.A. Millikan (1868-1953): It has been convincingly demon started that the electron was a particle charge. We will find, as our study progresses, that the electron is the principle character in electronics. The mass and charge of the electron were not known individually until 1909, when classical experiment, Millikan Oil Drop Experiment, was performed by R.A. Millikan. This experiment yielded a numerical value for the charge of the electron, and this charge was shown to be the same as that predicted earlier by Stoney. With both the charge to mass ratio and the charge known, it was possible to calculate the mass of the electron.

12.15 Bell Telephone Laboratories (1925–1984): The point-contact transistor, which was created at Bell Laboratories in 1947 by John Bardeen, Walter Brattain, and William Shockley, was the first functional transistor. The first widespread application of transistors occurred in the early 1950s when Shockley's improved bipolar junction transistor went into production in 1948.

Nokia Bell Labs is an American industrial research and scientific development business owned by the Finnish company Nokia. It was formerly known as Bell Telephone Laboratories (1925-1984), then AT&T Bell Laboratories (1984-1996), and Bell Labs Innovations (1996-2007). With its main office located in Murray Hill, New Jersey, the company runs a worldwide network of laboratories.

The development of the transistor, laser, photovoltaic cell, charge-coupled device (CCD), information theory, the Unix operating system, programming languages B, C, C++, S, SNOBOL, AWK, AMPL, and others is attributed to researchers at Bell Laboratories. Bell Laboratories has produced ten Nobel Prize winners.

1.3 ELECTRICAL CURRENT

An electric current is a flow of charge, the unit of current is expressed in terms of a given quantity of charge flowing past a given point in a given length of time. We could define the unit of current to be one electron charge per second. Instead we choose the Coulomb as the unit of charge and the second as the unit of time and define as our unit of current as the Coulomb per second. We attach the name ampere to this unit of current. Thus

1 Ampere = 1 coulomb/ second
1 Ampere = 1 ampere (1 amp)
10^{-3} Ampere = 0.001 ampere = milliampere (1mA)
10^{-6} Ampere = 0.000001 ampere = 1 micro ampere (1 μA)
10^{-9} Ampere = 0.0000000001 ampere = 1 milli- micro ampere (1 Nano Ampere)
10^{-12} Ampere = 0.000000000001 ampere = 1 micro- micro ampere (1 Pico Ampere)

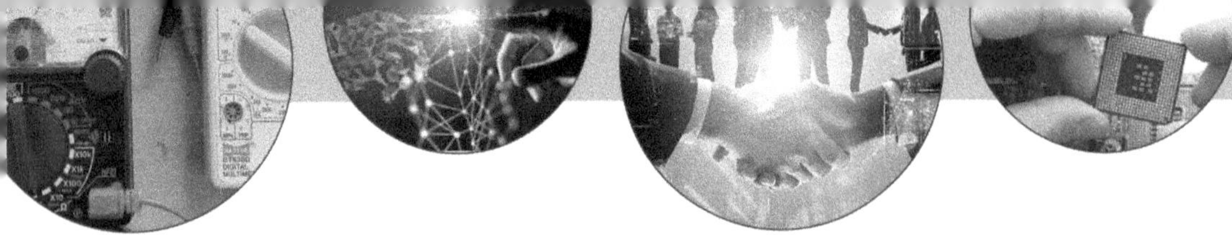

Since the electrons in material bodies are freer to move about than the protons, it would seem logical to define the direction of current to be the direction of electron flow. This, however would conflict with tradition, so we will define the direction of the conventional current to be the direction of flow of positive charge.

In most of the circuits and devices of electronics, the current consists of the electrons moving in a direction opposite to that of the conventional current. Confusion arising from this fact can be avoided by understanding that ignoring the atomic scale of the current process, a flow of positive charge in one direction is exactly equivalent to a flow negative charge in the other.

1.4 APPLICATIONS OF ELECTRONICS

Following table shows the commercial use of electronics

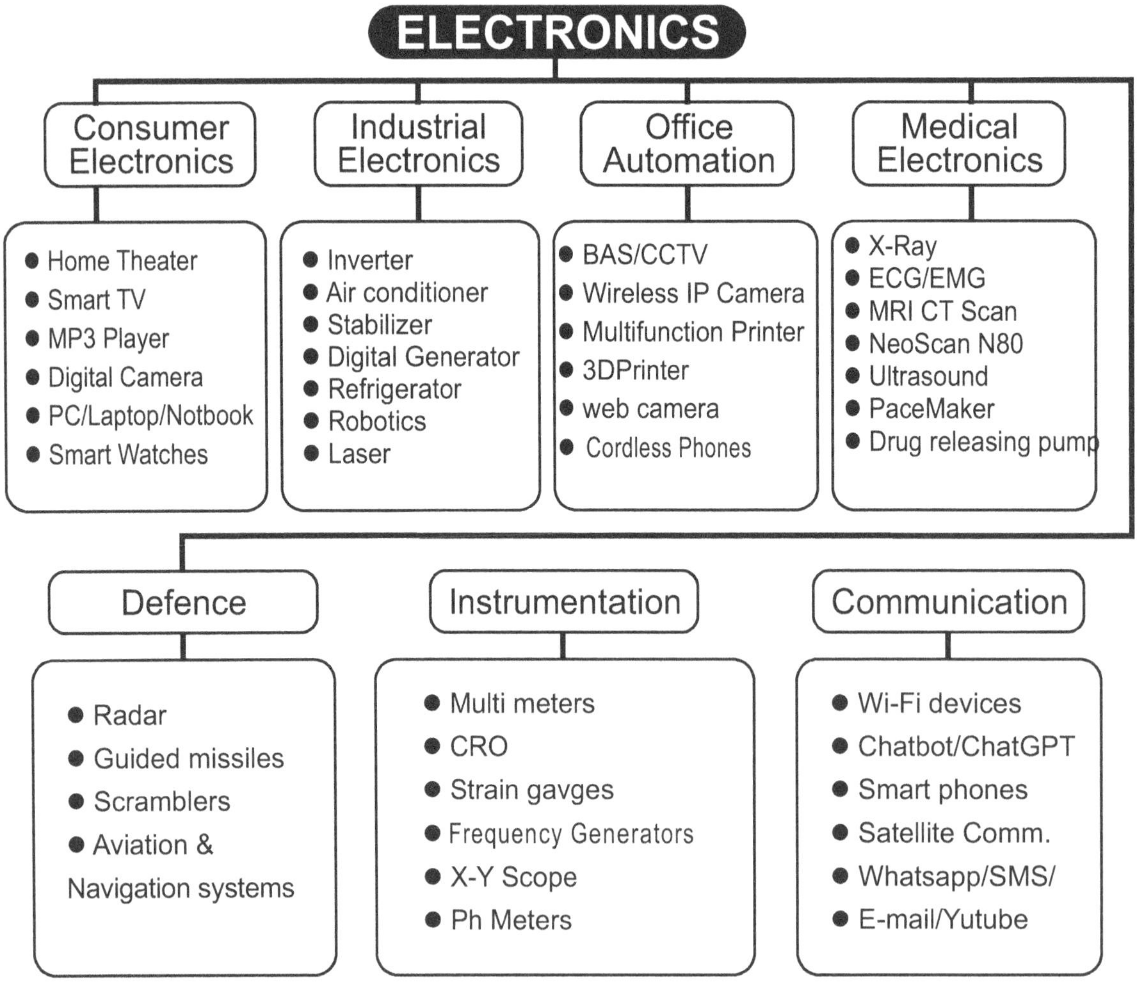

CHAPTER-2

Introduction to Electronics

Electricity comes to our homes and offices by the movement of electrons through wires, i.e. electric current is the movement of electrons. To understand the electronics we have to define nature of electrons atom can be broken up into smaller particles (electrons, protons, neutrons). The question is how charge flows? In order to understand this, we have to look inside an atom (atoms combine and forms matter.) All the atoms consist of nucleus and few electrons which revolve around nucleus at very high speed. Off all these electrons some electrons could be disturbed very easily, these are called as free electrons.

2.1 CONCEPT OF ELECTRONS

The Greek word for amber is where the term "electron" originates. In order to understand electrical processes, this material was crucial. For instance, the ancient Greeks were aware that rubbing fur on amber might leave an electric charge on the material's surface, which could then result in sparks.

While researching cathode ray tubes in the Cavendish Laboratory at Cambridge University in 1897, J.J. Thomson made the discovery that the electron was a subatomic particle. A sealed glass cylinder with two electrodes separated by a vacuum is known as a cathode ray tube. The tube glows as a result of cathode rays that are produced when a voltage is placed across the electrodes.

Thomson carried out experiments to determine that an electric field could deflect the rays and that the negative charge could not be removed from the beams (by the use of magnetism). He came to the conclusion that these rays were made up of negatively charged particles he dubbed "corpuscles," not waves.

The fact that their mass-to-charge ratio was over a thousand times lower than that of a hydrogen ion, as determined by his measurements, indicated that they were either extremely strongly charged or very little in mass. The latter conclusion was supported by further research by other scientists. The cathode material and initial gas in the vacuum tube had no effect on their mass-to-charge ratio. Thomson came to the conclusion that they applied to all materials as a result.

An individual atom's electrons do not all orbit the nucleus in the same plane. A metal becoming so strongly magnetised that all of its atoms have their rotations synchronised, as is the case in this fig. 2.1, is likewise extremely uncommon.

Scientists conducted numerous tests to determine the direction of flow of electricity in circuits once it was discovered, but in those early stages they were unable to do so.

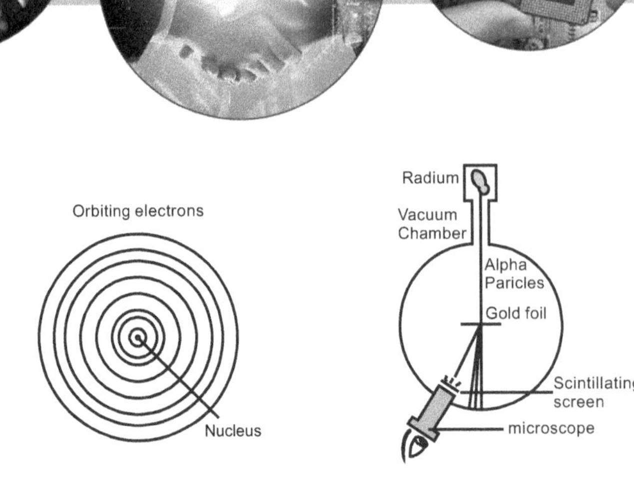

Fig. 2.1 : The planetary model of the atom and Rutherford's apparatus.

As they were aware that there were two different forms of electric charge—positive (+) and negative (-)—they chose to define electricity as the transfer of positive charge from + to -. Although they were aware that this was just a hunch, they had to decide. If electricity flowed in the opposite direction, from - to +, it could also account for all that was understood at the time. Electricity flowing from + to - had already been established by the time the electron was found (conventional).

In electronics, we are mainly interested in the outer electron shell of the atom. The electrons in the outer shell are known as Valence electrons, and the outer shell is known as the Valence shell. It is interesting to note that the valence shell never has more than eight electrons. In fact the valence electrons are easily freed from the atom. In fact the valence electrons are easily freed from the atom. In fact both the electrical and chemical properties of the elements depend on the action of the valence electrons.

Elements that have their valence shells almost filled are very stable, although they are not as stable as those whose valence shells are completely filled. These substances seek out free electrons in an effort to fill their valence shells. Consequently, elements of this type have very few free electrons in their atomic structure.

Substances with five or more electrons are called as insulators, where as elements with three or less valence electron tend to go them up easily and hence they are conductors. Some electrons, such as Silicon and Germanium, have four electrons in their valence shell. These substances are referred regarded be semiconductors since they are neither good insulators nor conductors. Semiconductors are used extensively in electronics.

2.2 WHAT IS ELECTRONICS

Electronics is a branch of science which deals with the flow of electrons and it uses the property of matter under various conditions for generating meaningful devices. The first component of electronics was developed in the year 1906 and was named as triodes since it has three electrodes. Electronics deals in the micro & mille range of voltage, current and power to control kilo & mega volts, amperes and watts. In the present age of electronics you will find its application in homes, factories medical sciences, defence industries and everywhere. Basically we can divide all electronic components into two primary categories i.e. Active & Passive devices. Active devices are capable of generating or amplifying the energy where as passive device neither generate nor amplifies the energy. Resistors, capacitors & inductors are example of passive components and on the other hand battery, transistors etc. are active components.

2.3 ELECTRONS AND ELECTRICITY

Actually, the picture of electrons "bound" to atoms is really not correct for most solids, and particularly for metals. Why should not electrons be able to move wherever they please? It is a free country! And in metals, electrons really are "free" in the sense they basically spread throughout a very large area containing a very large number of atoms (the wave functions are "de-localized").

The issue with "shell configurations" is that as atoms grow closer together; their electron "orbits" begin to overlap. This results in "bands" of electron states, many of which are de-localized, rather than atomic shells. When you apply an electric field, the electrons acquire some momentum and their occupation of these "bands" sloshes in the direction of the electric field. Think of individual electrons in the metal as spread out over a quite large area, and then starting to move in response to the electric field. There is your electricity. The "jumping" concept (while still possible - through quantum tunnelling) applies only under special circumstances - most of the time we are interested, the electrons can just move continuously without any jumping.

2.4 ATOMIC NUMBER

The number of units of charge that were present in the nuclei of the various chemical elements was only very roughly known to scientists at the time They were unaware of the relationship between an element's location on the periodic table and its atomic number, which we now refer to as the number of nuclear charge units. Mendeleev's table didn't appear to have any essential physical meaning; rather, it only seemed to be a tool for organisation. It was unclear how to number the elements progressively using integers because everything Mendeleev had done appeared to hold true whether you turned the table upside-down or flipped its left and right sides. Contrary to the contemporary table, Mendeleev's original was actually backwards.

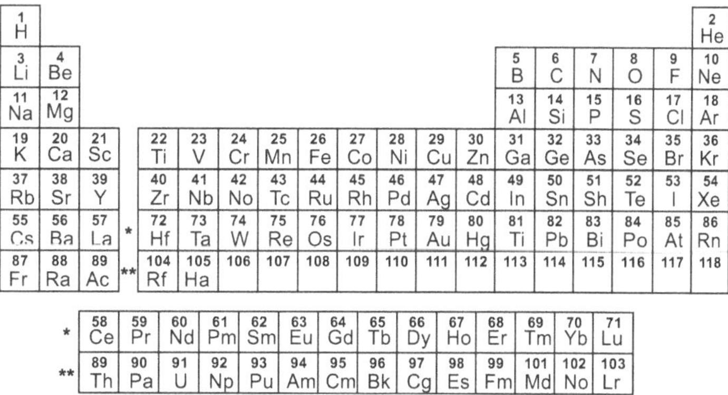

1 H																	2 He
3 Li	4 Be											5 B	6 C	7 N	8 O	9 F	10 Ne
11 Na	12 Mg											13 Al	14 Si	15 P	16 S	17 Cl	18 Ar
19 K	20 Ca	21 Sc	22 Ti	23 V	24 Cr	25 Mn	26 Fe	27 Co	28 Ni	29 Cu	30 Zn	31 Ga	32 Ge	33 As	34 Se	35 Br	36 Kr
37 Rb	38 Sr	39 Y	40 Zr	41 Nb	42 No	43 Tc	44 Ru	45 Rh	46 Pd	47 Ag	48 Cd	49 In	50 Sn	51 Sh	52 Te	53 I	54 Xe
55 Cs	56 Ba	57 La *	72 Hf	73 Ta	74 W	75 Re	76 Os	77 Ir	78 Pt	79 Au	80 Hg	81 Ti	82 Pb	83 Bi	84 Po	85 At	86 Rn
87 Fr	88 Ra	89 Ac **	104 Rf	105 Ha	106	107	108	109	110	111	112	113	114	115	116	117	118

	58 Ce	59 Pr	60 Nd	61 Pm	62 Sm	63 Eu	64 Gd	65 Tb	66 Dy	67 Ho	68 Er	69 Tm	70 Yb	71 Lu
*	58 Ce	59 Pr	60 Nd	61 Pm	62 Sm	63 Eu	64 Gd	65 Tb	66 Dy	67 Ho	68 Er	69 Tm	70 Yb	71 Lu
**	89 Th	90 Pa	91 U	92 Np	93 Pu	94 Am	95 Cm	96 Bk	97 Cg	98 Es	99 Fm	101 Md	102 No	103 Lr

Fig.2.2 : A modern periodic table, labeled with atomic numbers.

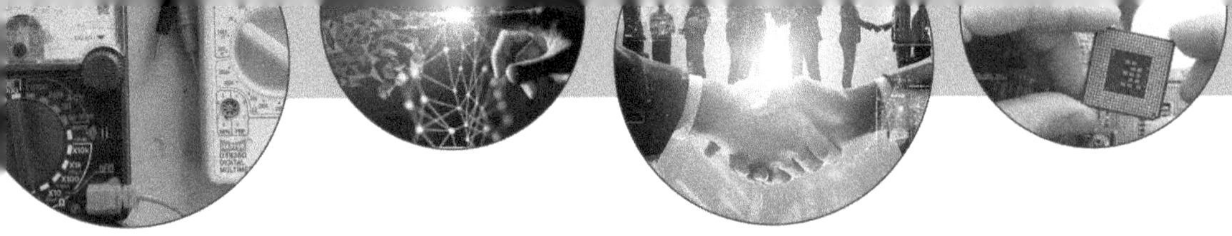

2.4.1 The Proton

Many physicists hypothesized that the nucleus might include smaller particles with distinct charges of +e rather than being a point-like object because the nuclear charges were all integer multiples of e. It didn't take long for this theory to be proven correct. Rutherford reasoned that the modest charge of the target nuclei would result in a very mild repulsion if he used alpha particles to attack the atoms of a very light element. It's possible that the few alpha particles that did arrive on head-on collision paths might get so near to the target nuclei that they would actually strike them. Since an alpha particle is a nucleus in and of itself, there would be a violent collision between two nuclei as a result of the high speeds involved.

In an experiment involving alpha particles striking a target containing nitrogen atoms, Rutherford struck gold when it was discovered that the target was producing charged particles that were seen to be blasting out like car components after a quick collision. The deflection of these particles in electric and magnetic fields was measured, and the results revealed that they had the same charge-to-mass ratio as hydrogen atoms that had been singly ionised. These hypothetical single-charged particles, which were ultimately given the name protons, were identified by Rutherford as the nucleus' charged particles. The hydrogen atom has one proton in its nucleus, and generally speaking, the atomic number of an element indicates how many protons are present in each of its nuclei. The proton's mass is nearly 1800 times larger than the electron's mass.

2.4.2 The Neutron

If all the nuclei could have been created using only protons, it would have been lovely and easy, but that is not possible. A periodic table will quickly reveal that while many of the atomic masses are not exactly integer multiples of the mass of hydrogen, many others are very close so. The mass of an element other than hydrogen is always more than its atomic number, not equal to it, even when the masses are close to whole numbers For instance; helium contains two protons but a mass that is four times larger than hydrogen's.

Chadwick cleared up the confusion by proving the existence of a new subatomic particle. He gave it the name neutron because it is electrically neutral, in contrast to the electron and proton, which are both charged particles. The approach of Chadwick's experiment was to expose a sample of the light element beryllium to a stream of alpha particles from a lump of radium. This experiment is detailed in chapter 4 of book 2 of this series. Because there are just four protons in beryllium, an incoming alpha can actually collide with the nucleus rather than being deflected away from it by electrical repulsion. Neutrons were observed as a new form of

	charge	Mass in Units of the Proton's		Location in
		Mass	Atom	
Proton	+e	1		in nucleus
Neutron	0	1.001		in nucleus
Electron	-e	1/1836	nucleus	orbiting

radiation emerging from the collisions, unsuspected components of the nucleus that had been knocked out. Chadwick also discovered the mass of the neutron, which is roughly identical to that of the proton, as stated in Conservation Laws.

Neutrons provided an explanation for the enigmatic elemental masses. For example, the mass of helium is very nearly four times more than that of hydrogen. This is due to the fact that in addition to its two protons, it also has two neutrons. An atom's mass is mostly determined by the total amount of protons and neutrons in the atom. The atom's mass number is consequently defined as the sum of the neutron and proton numbers.

Physicists only had imprecise estimations of the charges of the various nuclei in the era immediately following the discovery of the nucleus. They simply determined the greatest number of electrons that could be removed from the lightest nuclei using various techniques, such as chemical reactions, electric sparks, ultraviolet radiation, and so on.

They could easily take one or two electrons out of helium to form He+ or He++, for example, but no one could create He+++, perhaps because helium only possessed a nuclear charge of +2e. Unfortunately, only a few of the lightest elements could be entirely stripped of their electrons because, as more electrons were removed, the positive net charge that remained increased and the remaining negatively charged electrons were more securely held on.

The atomic numbers of the heavy elements could only be loosely extrapolated from those of the light elements, where the atomic number was about equal to one-half of the mass of an atom expressed in units of hydrogen. For instance, the mass of gold was calculated to be around 197 times greater than that of hydrogen, or about 100 atomic units. Now we know it's 79.

To deflect through the same angle, an alpha particle needs to be significantly closer to the low-charged copper nucleus. In electronic circuits, resistors, capacitors, and inductors are the three fundamental parts.

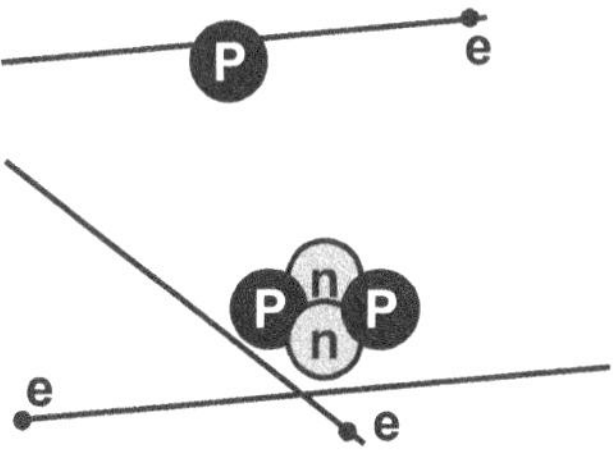

fig. 2.3 : Construction of atoms: hydrogen (top) and helium (bottom).

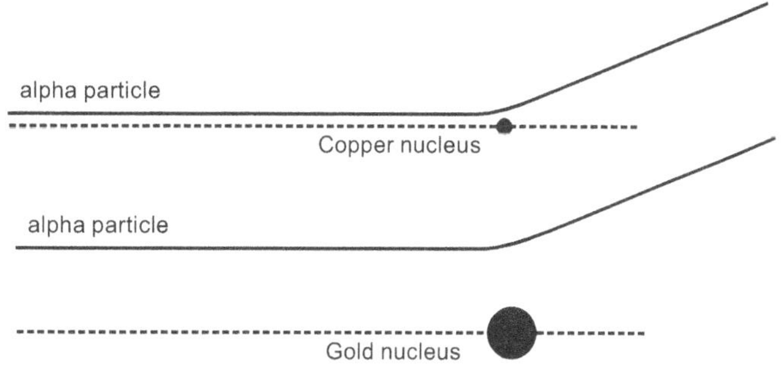

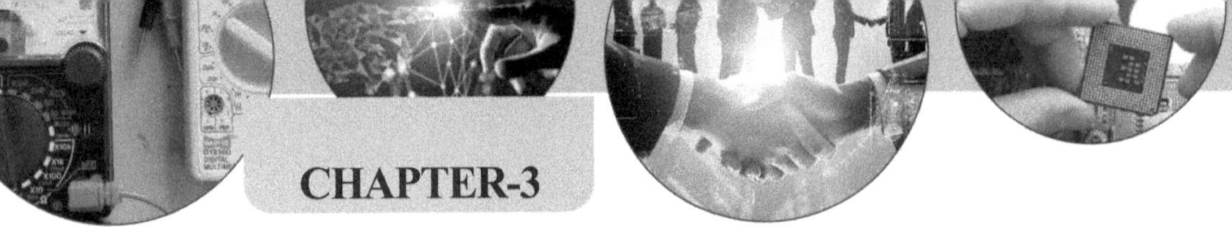

CHAPTER-3

Basic Components

3.1 RESISTORS

The characteristic of a component that **restricts the flow of electric current** is called resistance. The component generates heat as a result of energy being used up as the voltage across it drives the current through it.

Many times it becomes necessary to limit the flow of current through a circuit. The component used to perform this job is called as resistance. The physical appearance used to represent are as shown below :

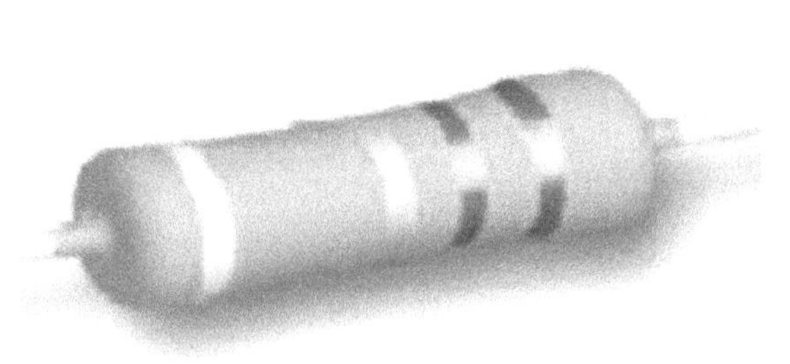

Fixed resistance

Variable resistance

Ω are a unit of resistance; an omega represents an ohm.

Resistence are frequently stated in kΩ and MΩ because 1 is a very modest number for electronics.

$1\,K = 1000\Omega$, $1\,M = 1000000\Omega$; K-Kilo ohm , M- Mega ohm

Electronics resistors come in a variety of resistances, ranging from 0.1 ohms to 10 MΩ.

3.1.1 Function

Electric current flow is restricted by resistors; for instance, a resistor connected in series with an LED can reduce the current that passes through the LED.

Resistor values - with help of resistor colour code

Resistor values are normally shown using coloured bands. According to the table, each colour denotes a different number. Most resistors have 4 bands:

The first band gives the first digit.

The second band gives the second digit.

The number of zeros is displayed in the third band.

The tolerance is displayed using the fourth band.

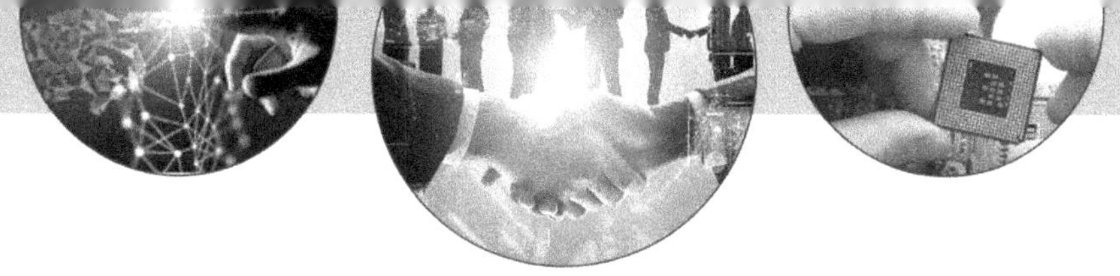

The Resistor Colour Code	
Colour	**Number**
Black	0
Brown	1
Red	2
Orange	3
Yellow	4
Green	5
Blue	6
Violet	7
Grey	8
White	9

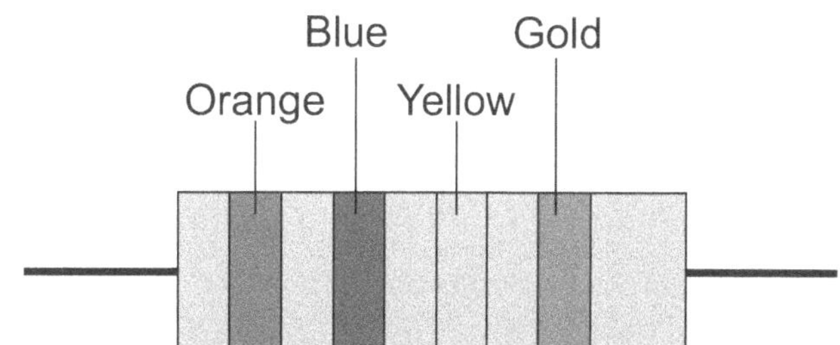

Fig.3.1: The Resistor Colour

In Fig 3.1,This resistor has Orange (3), Blue (6), yellow (4 zeros) and gold bands.
So its value is 360000 = 360 kilo Ohm

3.1.2   Tolerance

The resistor's tolerance, which is expressed as a percentage, is its accuracy. For example a 360 resistor with a tolerance of ±10% will have a value within 10% of 360, between 360 - 36 = 324 and 360 + 36 = 396 (36 is 10% of 360).

Gold ±5%, Silver ±10% Red ±2%, Brown ±1%.

The tolerance is ±20% if the fourth band is not visible. For practically all circuits, tolerance can be disregarded because accurate resistor values are rarely needed.

3.1.3   Power Ratings of Resistors

When current passes through a resistor, electrical energy is transformed to heat. The impact is usually insignificant, but if the voltage across the resistor is high or the resistance is low, a significant current may flow, causing the resistor to become visibly warm. Resistors have power ratings that demonstrate their ability to tolerate the heating effect.

Resistor power ratings are rarely mentioned in parts lists since 0.25W or 0.5W standard power ratings are sufficient for the majority of circuits. There will be circuits using low value resistors (less than roughly 300) or high voltages, which are the few instances where a higher power is required and should be clearly indicated in the parts list (more than 15V).

$P=V^2/R$, or $P=I^2 R$

P = watts of power generated inside the resistor (W)

I is the resistance's current in amps (A)

R is the resistor's ohmic resistance.

V = the volts across the resistor (V)

High power resistors 25W

Examples: A 360 resistor with 12V across it requires a

power rating of $P = V^2/R = 12^2/360 = 0.4W$.

High power resistors 5W

A typical 0.5W resistor would be appropriate in this situation. A 36Ω resistor with 12V across it requires a power rating of P = V²/R = 12²/36=4.0W. A 5W high power resistor would work well.

Note: The E 96 series of resistors has six bands-the first four for the value, the fifth for tolerance and the sixth for temperature stability.

Resistance of any conductor depends on three factors, length of the conductor (L),area of cross section (A) and resistivity of that material (also called as specific resistance (p) The relationship between these factors with resistance

R=pL/A

Resistors can be grouped in two categories i.e., fixed and variable. In the case of fixed resistor, you can not change its value, whereas in the case of variable resistor, you can set its value at any point between its lower & upper limits. Fixed resistors can be wire wound carbon film type or deposited film type as shown in fig. Out of these three types, wire wound resistors are more precise in value.

3.1.4 Variable Resistors

Variable resistors are used to control volume in Radio/T.V., brightness etc. Their values can be changed in one way or the other. Fig. 3.2 illustrates the construction of variable resistors.

In addition to being used as a potentiometer with all three connections, variable resistors can also be used as a rheostat with just two connections (the wiper and one end of the track). For configuring circuits that won't need additional tweaking, presets, or miniature versions, are created.

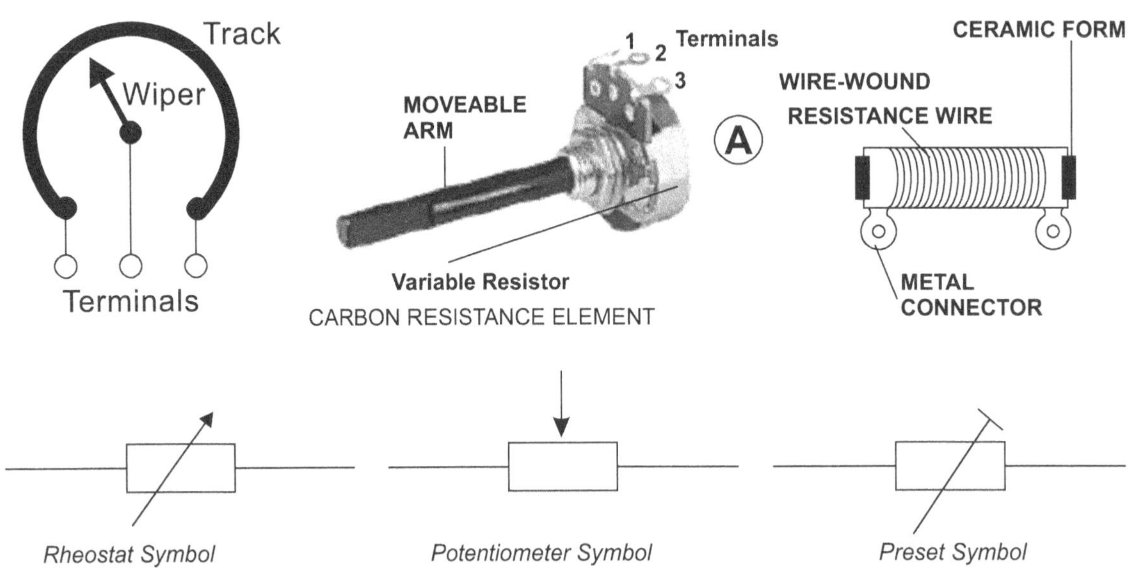

Fig.3.2 : Variable Resistors & Symbols

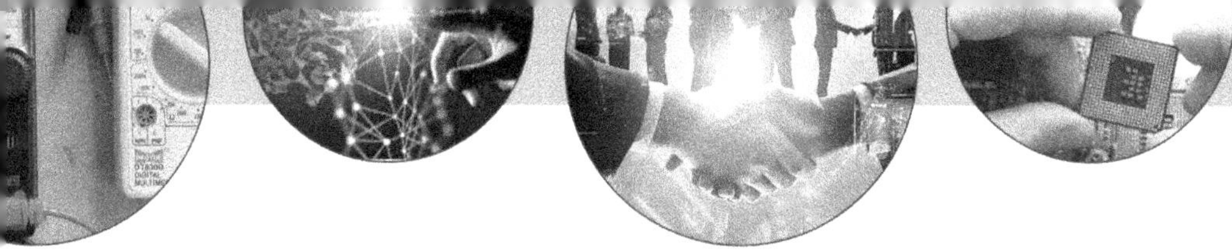

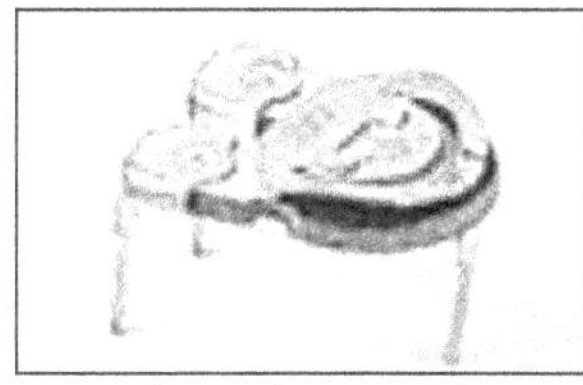
Preset (open style)

Preset (open style)

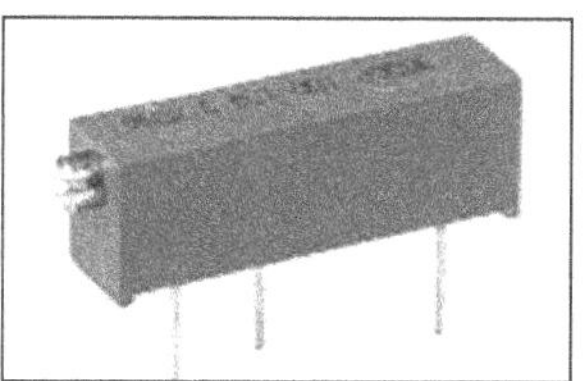
Preset (open style)

Fig.3.3 : Physical Appearance of Variable resistors

Rheostat

The easiest way to use a variable resistor is with a rheostat. Fig. 3.3 The track's end and the moveable wiper are linked to one of the two terminals. The resistance between the two terminals is changed by rotating the spindle from zero to the highest resistance.

The brightness of a lamp or the speed at which a capacitor charges are two examples of how rheostats are frequently used to change current.

If the rheostat is mounted on a printed circuit board, then all three connections may be connected. One of these will, however, be connected to the wiper terminal. However, it serves no electrical purpose and just improves the mechanical stability of the attachment.

3.1.5 Potentiometer

Potentiometers that use variable resistors have three linked terminals.

This configuration is typically used to change the voltage, for instance, to determine the sensor-based circuit's switching point or to regulate the volume (loudness) in an amplifier circuit. The wiper terminal will offer a voltage that can be adjusted from zero to the supply's maximum if the terminals at the track's ends are linked across it.

Presets

The variable resistors in question are tiny copies of the common type. They are made to be put right into the circuit board and modified only after the circuit is constructed. Set the sensitivity of a light-sensitive circuit, for instance, or the frequency of an alert tone. Presets must be adjusted using a tiny screwdriver or equivalent instrument.

Presets are used in projects occasionally in place of ordinary variable resistors because they are significantly less expensive than those resistors. Multiturn presets are employed when extremely precise adjustments are required. The slider can be moved from one end of the track to the other by turning the screw at least ten times, which provides extremely fine control.

Following table gives details about characteristic of various types of resistors.

Property Wirewound	Carbon Type	Carbon composition	Metal film	oxide
Maximum value	20 MW	10 W	100 W	270 W
Tolerance	+ 10%	+5%	+2%	+ 5%
Power rating	0.125-1 W	0.25-2W	0.5 W	2.5 W
Stability	Poor	Good	Very good	Very good
Use general	General	General	accurate work	low values

Figure shows the rear-view of the inside of variable carbon resistor. The resistance between terminals 1 and 2 rises when the shaft is turned in the direction indicated by the arrow, while the resistance between terminals 3 and 4 decreases. between terminal 2 and 3 decreases. This is because the length of the resistive material between terminals 1 and 2 increases, while the length between terminals 2 and 3 decreases. A potentiometer, commonly referred to as a pot, is a form of variable resistor. Three terminals make up a potentiometer.

3.1.6 Light Dependent Resistor (LDR)

Besides these, some special purpose resistors are also used in specific applications, e.g. LDR (Light dependent resistor), value of resistance changes with the intensity of light, it is maximum when dark & reduces with increase in intensity of light. This is used in the case of light operated switch, object counter. Cadmium sulphide (CdS) is used to make it.

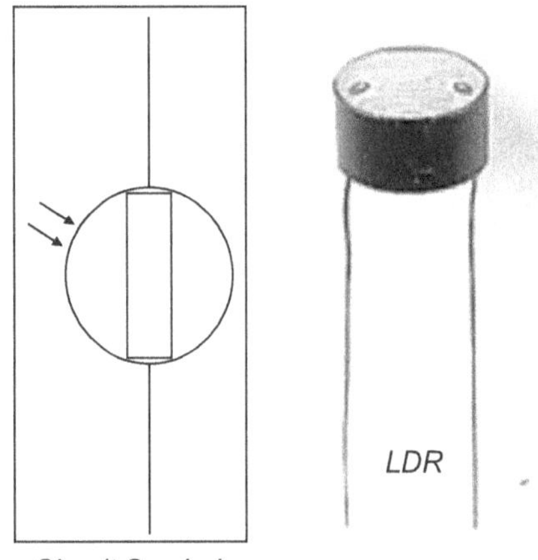

For a standard LDR typical results, we can measure using a multimeter to measure the resistance Darkness: maximum resistance, about 1M. Very bright light: minimum resistance, about $100\,\Omega$

3.1.7 Thermistor

An input transducer (sensor) called a thermistor transforms temperature (heat) to resistance. The resistance of almost all thermistors lowers as temperature rises due to the negative temperature coefficient (NTC) of these devices.

3.1.8 Connecting Resistors

The resistors can be connected in two ways i.e. series and parallel. If first leg of first resistor is connected to the first leg of other resistor and second leg of first resistor is connected to the second leg of other resistor then the connection is called as parallel connection else it is series connection, as shown in fig.3.4 a &b.

Series: When resistors are connected in series, the total resistance is equal to the total resistance of each resistor. The combined resistance, R, of two resistors, R_1 and R_2, for instance, is given by:

Keep in mind that any resistance in series will always be larger than any resistance acting alone.

You may extend this to include more resistors by writing $R = R_1 + R_2 + R_3 + R_4 + ..R_n$

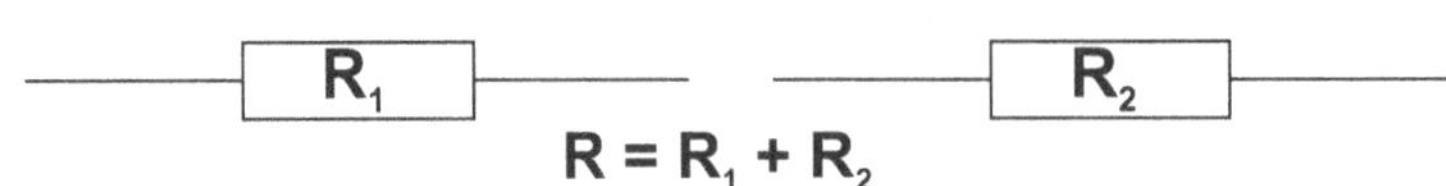

Fig.3.4a : Combining resistors in series

Parallel

Parallel resistor connections result in a combined resistance that is lower than each resistor's individual resistance. For example if resistors R_1 and R_2 are connected in parallel their combined resistance, R, is given by:

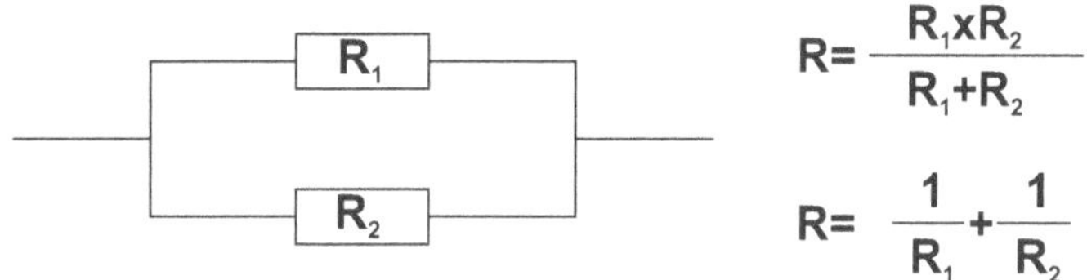

$$R= \frac{R_1 \times R_2}{R_1+R_2}$$

$$R= \frac{1}{R_1}+\frac{1}{R_2}$$

Fig.3.4b : Combining resistors in parallel

For Example:

EX.1. Calculate the circuit's overall resistance.

Solution:

Above two resistance 30 ohm, 30 ohm are in series, so the total resistance will be:

$R=R1+R2 \quad R = 30+30$

Total Resistance = 60 ohms

EX. 2. Calculate the circuit's overall resistance.

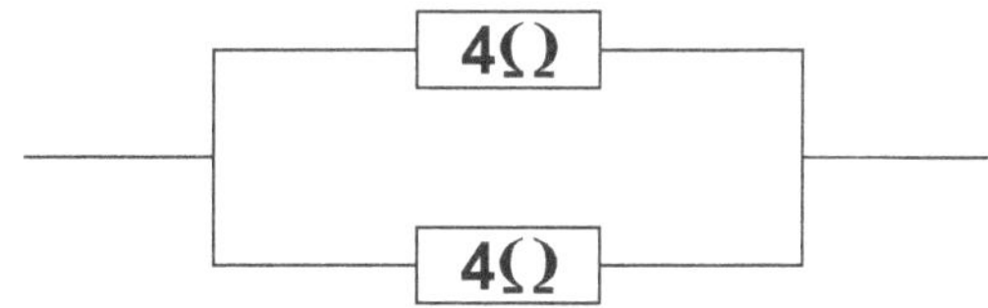

Solution:

Above two resistance 4 ohm, 4 Ω are in parallel, so the total resistance will be:

$R_1 = 4\,\Omega, \quad R_2 = 4\,\Omega$

$$R= \frac{R_1 \times R_2}{R_1+R_2}$$

$$\frac{1}{R}=\frac{1}{R_1}+\frac{1}{R_2} \quad \frac{1}{R}=\frac{1}{4}+\frac{1}{4} \quad R=\frac{4 \times 4}{4+4}=\frac{16}{8}=2.0\,\Omega$$

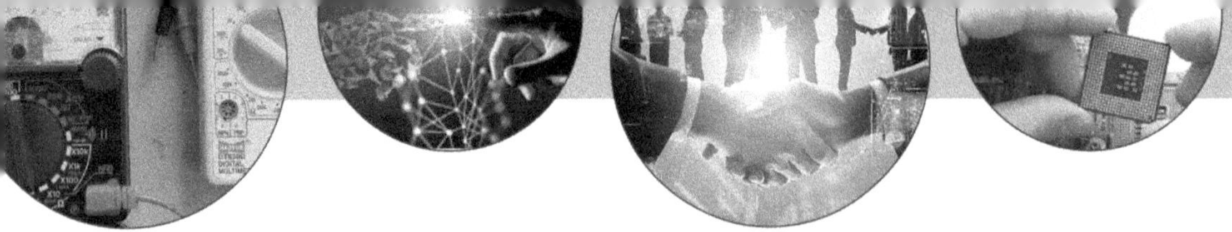

3.1.9 Measuring & Testing Resistor

Beside printed colour code, you can measure the value of resistor using ohm meter or multi meter. Many times resistors also go bad, common faults related to resistors are:

Open - Ohm meter will show infinite resistance.

Shorted - Ohm meter will show zero or very low resistance.

In order to test a resistor, always take it out of the circuit and then test the same. If this is not possible, then ensure that all power to the circuit is disconnected before connecting an ohmmeter.

Resistance of Conductors, Semiconductors and Insulators

An object's resistance is influenced by its shape and the material it is constructed of. Objects having a smaller cross-section or a longer length will have more resistance for a given substance.

Materials can be divided into three groups:

1. Conductors which have low resistance.

    Example: Metals (aluminum, copper, silver etc.) and carbon. Connecting wires,

    switch contacts, and lamp filaments are all made of metal. Long coils of thin wire or carbon are used to make resistors.

2. Semiconductors which have moderate resistance.

    Example: Germanium, silicon. Transistors, LEDs, diodes, and integrated circuits are

    all made from semiconductors (chips).

3. Insulators which have high resistance.

    Typical materials include paper, glass, and plastics like PVC (polyvinyl chloride) and polythene. In order to keep wires from touching each other, PVC is employed as their exterior covering.

3.2 INDUCTORS

Because of its inductance quality, an inductor is a passive electrical component used in electrical circuits. An inductor can take many forms.

A current-carrying conductor's magnetic field creates an effect known as inductance, which is measured in henries. A magnetic flux proportionate to the electrical current flowing through the conductor is produced. An electromotive force (emf) is a force that opposes a change in current by causing a change in magnetic flux, which is itself caused by a change in current. The generated emf for a unit change in current is measured by inductance. An inductor with an inductance of 1 Henry, for instance, generates an emf of 1 V when the current through it varies at a rate of 1 amp per second. The area of each loop or turn, how many turns there are, and what it is wrapped around all have an impact on inductance. By coiling the conductor around a material with a high permeability, for instance, the magnetic flux connecting these turns can be made stronger.

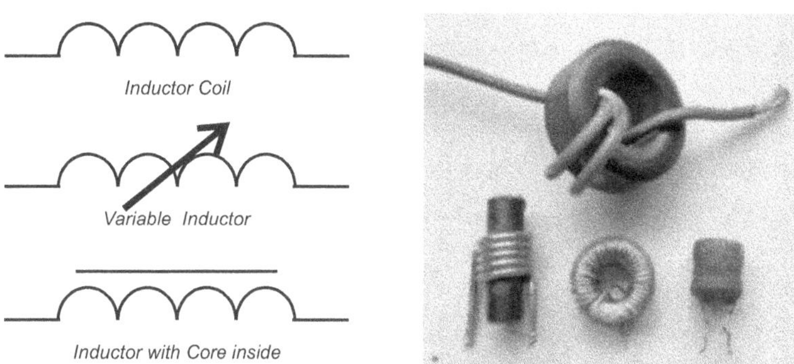

Fig. 3.5 : Inductors symbols & Physical appearance

3.2.1 Parallel Circuits

A parallel design of inductors results in the same potential difference between each one (voltage). The formula is same as used to determine resistance in parallel. To find their total equivalent inductance (Leq): as in fig. 3.6

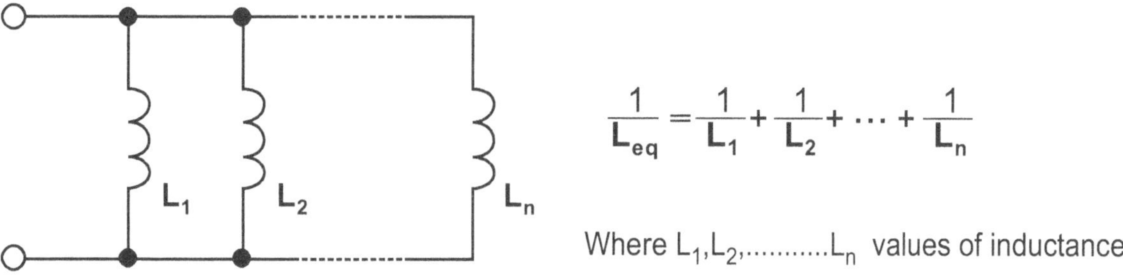

$$\frac{1}{L_{eq}} = \frac{1}{L_1} + \frac{1}{L_2} + \cdots + \frac{1}{L_n}$$

Where $L_1, L_2, \ldots \ldots L_n$ values of inductance

Fig.3.6 : Inductors in parallel

3.2.2 Series Circuits

The voltage across each inductor might vary, yet the current flowing through inductors connected in series remains constant. The overall voltage is equal to the sum of the potential differences. To find their total inductance: as in fig. 3.7

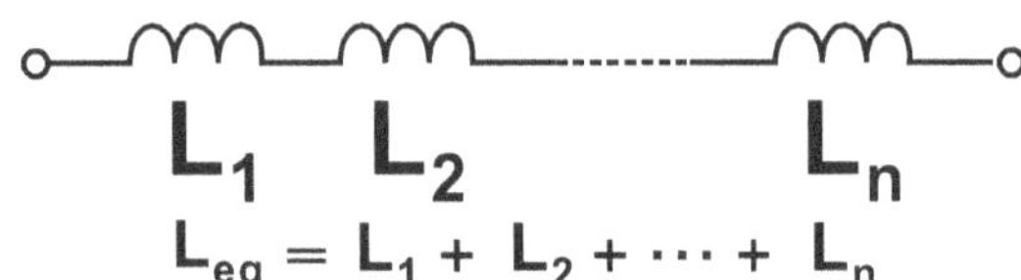

$$L_{eq} = L_1 + L_2 + \cdots + L_n$$

Fig.3.7 : Inductors in series

When magnetic fields between individual inductors are not mutually coupled, these straightforward relationships are valid. All inductors can be categorised into fixed or variable type.

In general the inductance of a core increase with, No. of turns Induction of core of magnetic material

3.2.3 Applications

In analogue circuits and signal processing, inductors are frequently utilised. Tuned circuits that may accentuate or filter out particular signal frequencies are made up of inductors, capacitors, and other components. This can range from the now-obsolete use of large inductors as chokes in power supplies that work in tandem with filter capacitors to remove lingering hum or other fluctuations from the direct current output to smaller inductances like those produced by a ferrite bead or torus wrapped around a cable to block the transmission of radio frequency interference. Tuned circuits are made possible by smaller inductor/capacitor combinations and are utilised, for example, in radio reception and broadcasting.

3.3 CAPACITORS

A capacitor is an electrical apparatus that may store energy in the electric field between two closely spaced conductors (referred to as "plates" in fig. 3.8). Electric charges of opposing polarity and equal magnitude accumulate on each plate of the capacitor when electricity is applied to it.

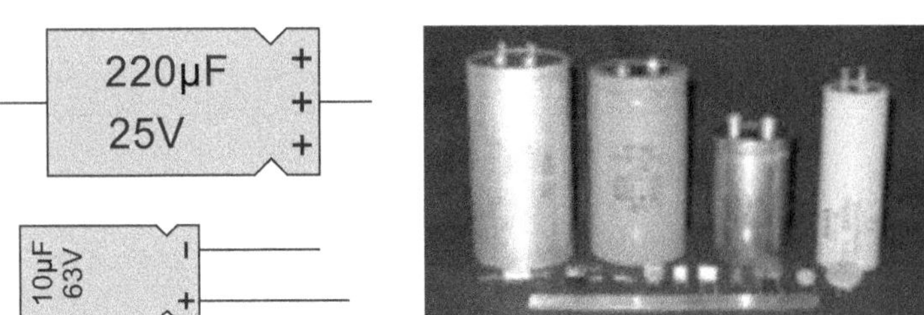

Fig.3.8 Various Capacitors Physical appearance

3.3.1 Capacitance

As seen in fig. 3.9, a capacitor is made up of two conductive electrodes, or plates, separated by a dielectric.

This electric field causes the plates of this straightforward parallel-plate capacitor to have a potential difference, $V = Ed$.

According to Fig. 3.9, the capacitance (C) of the capacitor is a measurement of the amount of charge (Q) stored on each plate for a specific potential difference or voltage (V) that emerges between the plates (I)

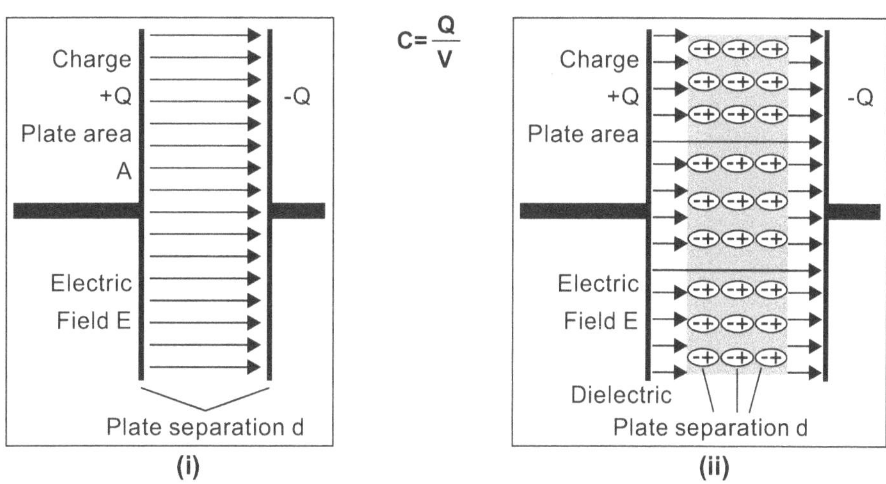

Fig.3.9 : Capacitance and electrical circuit

One coulomb of charge can be held in a capacitor thanks to a one volt applied potential difference across the plates, giving it a capacitance of one farad in SI units. Due to the farad's enormous size, capacitor values are typically represented in microfarads (μF), nanofarads (nF), or picofarads (pF).

The capacitance is inversely related to the separation between the plates and directly proportional to the surface area of the conducting plate. It is also proportional to the permitivity of the dielectric substance, which separates the plates and is a non-conducting material.

$$C \approx \frac{\varepsilon A}{d} \; ; \; A \gg d^2$$

When A is the area of the plates, d is the distance between them, and is the permitivity of the dielectric (Dielectric constant), the capacitance of a parallel-plate capacitor may be calculated.

Dielectric polarizations, which is illustrated in Fig. 3.7(ii), is the process by which the rotating molecules produce an opposing electric field that partially cancels the field produced by the plates.

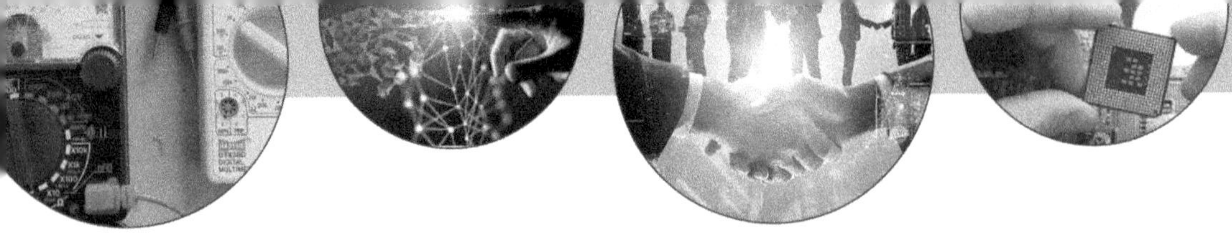

The electric field has an impact on the dielectric molecules' electrons, forcing the molecules to rotate just a little bit from their equilibrium locations. For the sake of clarity, the air gap is depicted; in a genuine capacitor, the dielectric makes direct contact with the plates. Moreover, capacitors restrict DC current while allowing AC current to flow.

This gauges how well a capacitor can hold onto electrical charge. More charge can be stored when the capacitance is high. The farad—symbol F—is used to measure capacitance. Prefixes are used to indicate the lesser values because 1F is a very large number.
There are four prefixes (multipliers) used: millifarad $(1mf)10^{-3}$, micro $(1\mu f)10^{-6}$, nano $(1nf)10^{-9}$ and pico $(1pf)10^{-12}$. stands for (millionth), thus $1000000F=1F$; n for (thousand-millionth); and $1000nF = 1F$. As p denotes (million-millionth), 1000pF equals 1nF.

Capacitors can be networked similarly to resistance but in the other direction, using series and parallel circuits to store the most energy possible.

3.3.2 Parallel Circuits

According to Fig. 3.10, parallel capacitors in a parallel setup have the same potential difference (voltage). The formula for their total capacitance (Ceq) is:

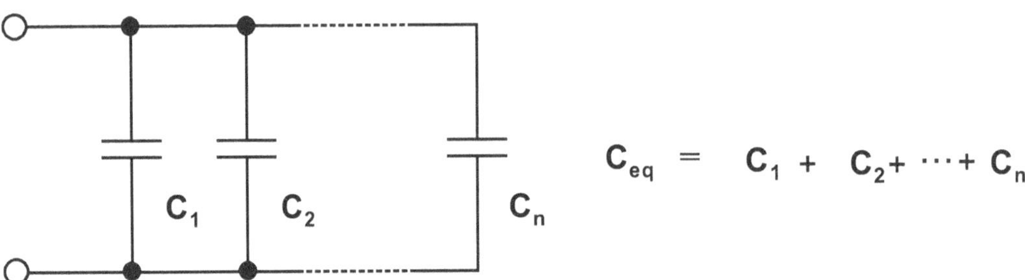

$$C_{eq} = C_1 + C_2 + \cdots + C_n$$

Fig.3.10 : Capacitance parallel circuit

To increase the overall quantity of charge held, capacitors are connected in parallel. In other words, increasing capacitance also results in a greater capacity to store energy. It reads as follows:

$$E_{stored} = \frac{1}{2}CV^2$$

3.3.3 Series Circuits

Although the voltage across each capacitor can vary, the current flowing through the series of capacitors in fig. 3.11 remains constant. The overall voltage is equal to the sum of the potential differences. Their combined capacitance is determined by:

$$\frac{1}{C_{eq}} = \frac{1}{C_1} + \frac{1}{C_2} + \cdots + \frac{1}{C_n}$$

Fig.3.11 : Capacitance series circuit

The entire capacitance has increased concurrently with the effective area of the combined capacitor. Because the plates are connected in series, the space between them has effectively grown, lowering the total capacitance.

In order to acquire very high voltage capacitors on a budget, capacitors will typically be connected in series, for example to smooth out ripples in a high voltage power supply. When three "600 volt maximum" capacitors are connected in series, their total working voltage rises to 1800 volts. Of course, the fact that the capacitance obtained is only one-third the value of the capacitors utilised offsets this. To combat this, three of these series setups can be connected in parallel, creating a 3x3 matrix of capacitors with the same total capacitance as a single capacitor but operable at three times the voltage. To ensure that the total voltage is distributed uniformly across each capacitor in this application and to discharge the capacitors for safety when the equipment is not in use, a big resistor would be connected across each capacitor.

The usage of polarised capacitors in alternating current circuits is another application; the capacitors are linked in series with reverse polarity so that at any given time only one of them is conducting.

3.3.4: Capacitor Types

Capacitor types come in a wide variety, however they can be divided into two groups: polarised (for large values, 1 microF or more) and unpolarized (for small values, up to 1microF). Each group's circuit symbol is unique.

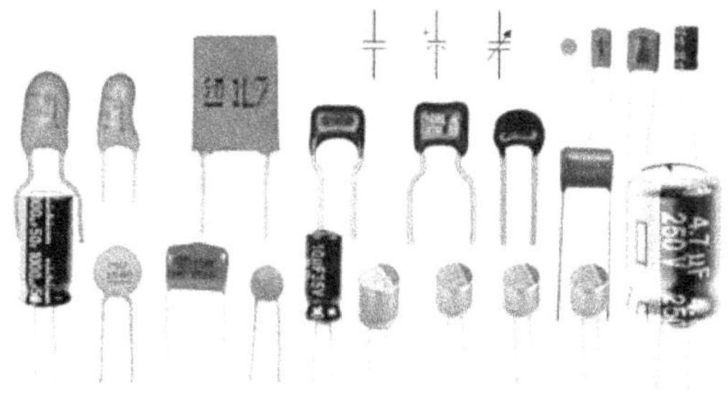

Fig.3.12 : Various Capacitors

Fig.: 3.12 shows various types of capacitors. In the sense of the watch rings from top to bottom, the following image displays several ceramic materials: multilayer ceramic, ceramic disc, multilayer polyester film, tubular ceramic, polyester, metalized polyester film, and aluminium electroplating.

Fig.: 3.12 shows Capacitors: SMD ceramic at top left; SMD tantalum at bottom left; through-hole tantalum at top right; through-hole electrolytic at bottom right.

Vacuum: two metals, usually copper, electrodes are separated by a vacuum A glass or ceramic envelope serves as the insulating material. They are typically employed in radio transmitters and other high voltage power devices because they have low capacitance (10–1000 pF) and high voltage (up to tens of kilovolts). Both fixed and variable types are available. Each tuned circuit can cover a whole decade of frequency using variable vacuum capacitors, which have a minimum to maximum capacitance ratio of up to 100. The most ideal dielectric, with a zero loss tangent, is vacuum. This enables the transmission of very high powers without significant loss or ensuing heating.

Air: Metal plates are separated by an air gap in air dielectric capacitors. The metal plates, which may be interspersed in large numbers, are often composed of silver-plated brass or aluminium. Radio tuning circuits use almost exclusively variable air dielectric capacitors.

Metallized Plastic Film: High-quality polymer films (often polycarbonate, polystyrene, polypropylene, polyester (Mylar), and polysulfide for high-quality capacitors) are combined with metal foil or a layer of metal placed on the surface to create metalized plastic film. They are suitable for timer circuits and have good quality and stability. Suitable for high frequencies.

Mica: Similar to metal film. Often high voltage. Suitable for high frequencies. Expensive. Excellent tolerance.

Paper: Used for relatively high voltages. Now obsolete.

Glass: Used for high voltages. Expensive. a stable temperature coefficient over a broad temperature range.

Ceramic: Alternating layers of metal and ceramic in the form of chips. The level of a device's dependence on temperature and capacity changes depending on whether it has a Class 1 or Class 2 dielectric. They frequently exhibit high frequency coefficients of dissipation, high dissipation factors (particularly for class 2), capacity that is dependent on applied voltage, and capacity that varies with age. In widespread low-precision coupling and filtering applications, they are widely used, though. Suitable for high frequencies.

Aluminum electrolytic: Polarized. Although the electrodes are built of etched aluminium to obtain significantly larger surfaces, they are constructed similarly to metal film. The dielectric is soaked with liquid electrolyte. They are capable of producing large capacities but have low tolerances, high instability, progressive capacity loss, especially when exposed to heat, and high leakage. Tend to lose capacity in low temperatures. They are unsuitable for high-frequency applications because to poor frequency characteristics. There are special varieties available with low equivalent series resistance.

Tantalum electrolytic: These capacitors are similar to aluminium electrolytic capacitors but have improved temperature and frequency properties. High dielectric absorption. High leakage. has significantly improved performance at low temperatures.

OSCON (or OS-CON) capacitors are a polymerized organic semiconductor solid

Electrolyte types with a longer lifespan but a higher price tag are available.

Super Capacitors: Solid-state polymerized organic semiconductor OSCON (or OS-CON) capacitors are electrolyte types with a longer lifespan but a higher price tag are available.Carbon aero gel, carbon nano tubes, or very porous electrode materials are used to make super capacitors. Very large capacity. can replace rechargeable batteries in some situations.

Gimmick capacitors are constructed by twisting together two insulated wires. Each wire forms a capacitor plate. Gimmick capacitors are also a sort of variable capacitor. Twisting and untwisting the two wires produces negligible (20% or less) variations in capacitance.

Specialized reverse-biased diodes with variable capacitance according to voltage are known as vertical capacitors. among other things, utilized in phase-locked loops.

According to the voltage level applied, a varactor is essentially a specialised reverse-biased diode employed as a variable capacitor.

3.3.5 Energy Stored (W)

A charged capacitors stores electrical energy. For charge (Q) and potential difference (V) the energy stored (W) is

$$W = \tfrac{1}{2}\,QV = \tfrac{1}{2}\,CV^2 \;(\text{since } Q = VC)$$

If Q is in coulombs and V is in volts, then W is in joules.

When a capacitor is not connected to its charging circuit, it can store electric energy and be used as a temporary battery. Electronic equipment frequently use capacitors to maintain power while batteries are being changed. (This stops information from being lost in volatile memory.)

In power supply, capacitors are used to smooth the output of a full or half wave rectifier. Moreover, they can be utilised in charge pump circuits as a form of energy storage to produce voltages higher than the input voltage.

Most electronic gadgets and bigger systems (like factories) use capacitors in parallel with their power circuits to shunt current variations from the main power source and provide a "clean" power supply for signal or control circuits. This technique is used, for instance, by audio equipment to shunt away power line hum before it enters the signal circuitry. Bypassing AC currents from the power source, the capacitors function as a local reserve for the DC power source. Automotive audio applications use stiffening capacitors to account for the resistance and inductance of the lead-acid car battery's leads.

3.3.6 Charging a Capacitor

In the circuit schematic, the capacitor I being charged by a supply voltage (Vs) while current is flowing through a resistor (R). Voltage measured across the The capacitor's initial voltage (Vc) is zero, but it increases as the capacitor charges. When Vc = Vs, the capacitor is fully charged. the determines the charging current (I) using the voltage a cross the resistor (Vs - Vc): Charging current,

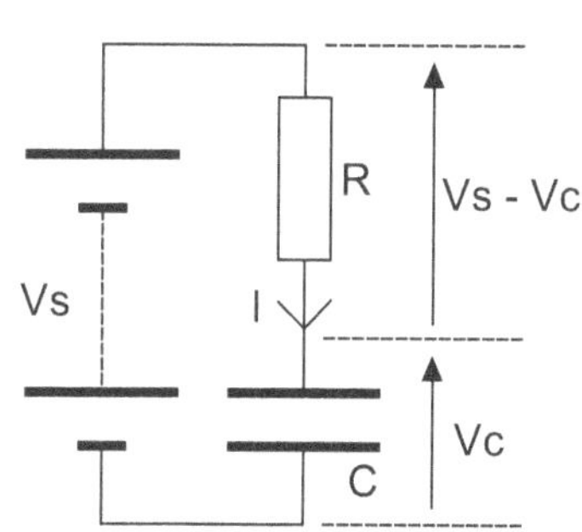

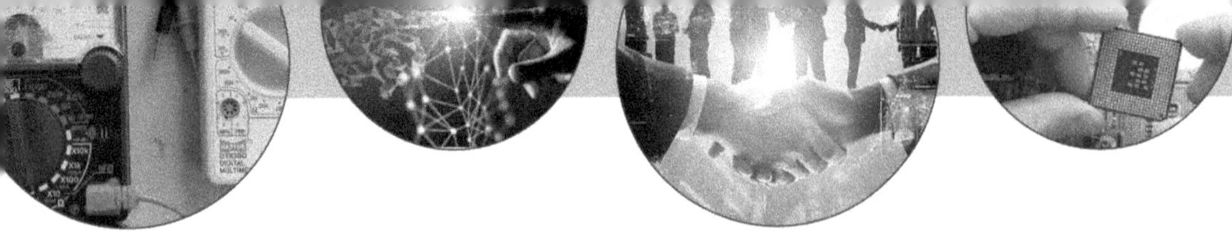

I = (Vs - Vc)/R (note that Vc is increasing) As soon as charge (Q) starts to build up (Vc = Q/C), the initial current, Io = Vs / R Vc, increases and, as a result, the voltage across the resistor and, consequently, the charging current, are reduced. As a result, the billing rate increases gradually.

$$\boxed{\text{TIME CONSTANT } = R \times C}$$

the time constant is in seconds where (s)

R represents resistance in ohms C is the capacitance in farads (F)

Consider this:

R = 47kΩ and C = 22°F result in a time constant of RC = 47k 22°F = 1.0 s.

If R = 33kΩ and C = 1F, the time constant RC = 33k1F = 33ms is produced. A high time constant indicates a slow capacitor charge. Be aware that the time constant is a characteristic of the circuit that includes the capacitance and resistance, not of a capacitor by itself.

3.3.7 Discharging a Capacitor

Capacitor should theoretically dissipate no power. It merely holds energy in reserve and releases it later. Although a perfect capacitor cannot be created, this condition can be approached. A capacitor is charged when it meets the applied voltage, as was previously stated. If the potential is not released, the capacitor will stay in this state for a while. Discharging the capacitor is the process of releasing the charge. Fig. 3.13 displays graphs of current and voltage during a capacitor discharging period.

$$\boxed{\text{CONSTANT} = RC}$$

The top graph demonstrates how the capacitor discharges when the current (I) diminishes. The initial voltage (Vo) across the capacitor and the resistance (R) define the initial current (Io): Io = Vo / R is the initial current.

Be aware that the current graphs depict a capacitor being discharged. A graph of this kind is a good illustration of exponential decay.

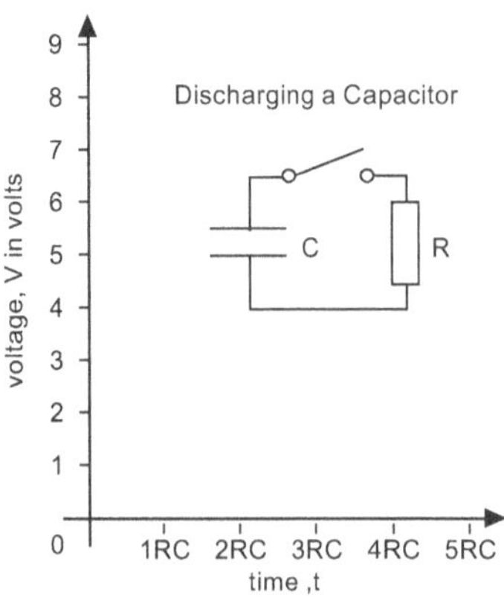

Fig.3.13 : Discharging Capacitors

3.3.8 Capacitive Reactance

A capacitor's resistance to AC is measured by its capacitive reactance (symbol Xc) (alternating current). Reactance is also measured in ohms, but it is more complex than resistance because the value of reactance depends on both the capacitance, C, and the frequency, f, of the electrical signal passing through the capacitor.

Capacitive reactance, $Xc = \dfrac{1}{2fC}$

where C is capacitance in farads, f is frequency in hertz (Hz), and Xc is reactance in ohms (F) At low frequencies, the reactance Xc is significant, whereas at high frequencies, it is modest. Because Xc is infinite for stable DC, which has zero frequency, capacitors pass AC but block DC, according to the rule.

3.3.9 Capacitor Number Code

On small capacitors, where printing is challenging, a number code is frequently used:

The first number is the value's first digit.

The second number is the value's second digit.

The third number is how many zeros are needed to represent capacitance as pF.

Any letters are merely for voltage rating and tolerance; ignore them.

These are a few instances:

(Not 102pF, but 1000pF = 1nF)

472J indicates (J signifies 5% tolerance) 4700pF = 4.7nF.

3.3.10 Capacitor Colour Code

Polyester capacitors have long been identified by a colour code. Although many of them are still in use, it is now obsolete. The top three colour bands, which represent the resistor code, should be read to get the value in pF. Disregard the tolerance and fifth bands (voltage rating). Due to the fact that there are no spaces between the colour bands, what appears to be two narrow bands are actually one wide band. The meaning of each colour for each of the bands is displayed in the table below.

Colour	Band 1 1st Digit	Band 2 2nd Digit	Band 3 Multiplier
Black	0	0	x 1
Brown	1	1	x 10
Red	2	2	x 100
Orange	3	3	x 1,000
Yellow	4	4	x 10,000
Green	5	5	x 100,000
Blue	6	6	x 1,000,000
Violet	7	7	-
Grey	8	8	-
White	9	9	-

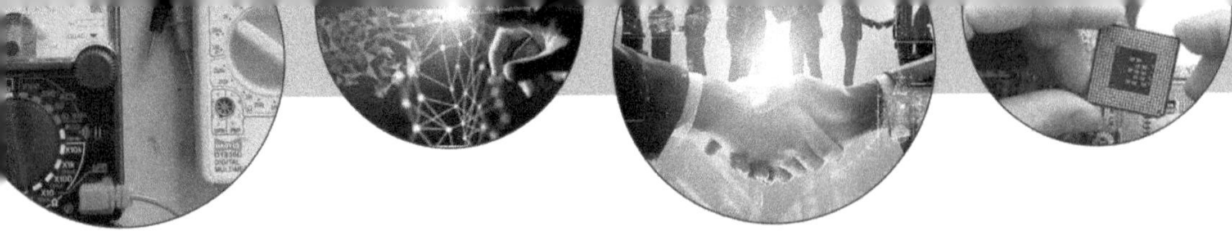

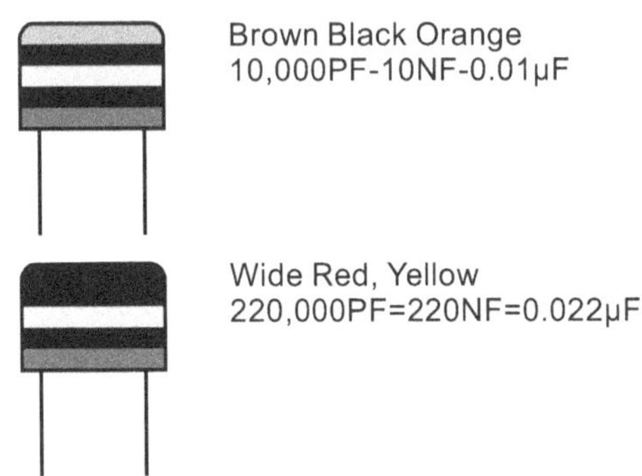

Above Figure shows a few examples of the capacitor colour code.

3.3.11 Uses of Capacitors

Many applications call for capacitors:

Timing, for instance, by using a 555 timer IC to regulate the charging and discharging.

Smoothing, such as in a power supply.

Coupling, for instance, connecting a loudspeaker and the various audio system stages.

Filtering, such as what happens in an audio system's tone control.

Tuning, like with radio equipment.

Energy storage, such as in the flash circuit of a camera.

3.3.12 Capacitor Coupling (CR-Coupling)

Due to the fact that capacitors pass AC (altering) signals while blocking DC (stable) signals, portions of electronic circuits may be connected by them. Capacitor coupling or CR-coupling is what this is. It is utilised between the audio system's stages to transmit the audio signal (AC) without any possible stable voltage (DC), such as when connecting a loudspeaker. The oscilloscope's 'AC' switch setting also uses it.

A capacitor coupling's time constant governs its precise behaviour (RC). Be aware that the resistance (R) might not be a distinct resistor but rather a component of the next circuit segment.

Signals must flow via a capacitor coupling in an audio system with minimal to no distortion. The lowest frequency audio signals required—typically 20Hz, T = 50ms—must have a time constant (RC) that is longer than the requisite time period (T).

Output when RC >> T

Since the time constant is far longer than the duration of the input signal, the capacitor does not have enough time to significantly charge or discharge, resulting in minimal distortion of the signal as it goes through.

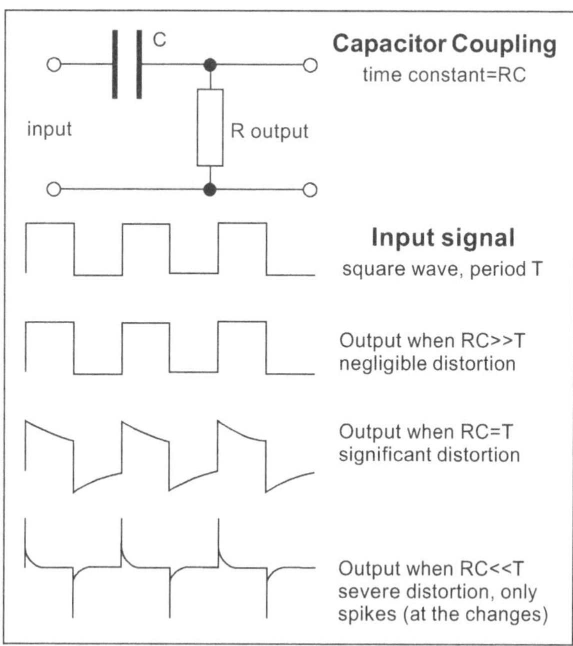

Output when RC = T

The capacitor has time to partially charge and discharge before the signal changes when the time constant is equal to the time period, as can be shown. As a result, the signal undergoes substantial distortion as it passes through the CR-coupling. See how the output is not affected by the input signal's abrupt changes because they pass directly through the capacitor.

Output When RC << T

With each abrupt change in the input signal, the capacitor has time to fully charge or discharge when the time constant is substantially lower than the time period. Indeed, only the abrupt changes make it to the output, where they show up as 'spikes,' alternately positive and negative. This can be helpful in a system that needs to overlook gradual changes in order to detect rapid changes in a signal.changes suddenly, but must ignore slow changes.

Capacitor	Polarized capacitors	Variable capacitor
	─)⊢+	
	─⊢+	
─⊢	─⊓+	─⊬

Capacitor symbols

CHAPTER-4

Electric Current

4.1 AN INTRODUCTION: Every aspect of our existence involves electricity. Our homes are illuminated by electricity, which also heats our meals and runs our computers, televisions, and other electronic equipment. Our autos run on electricity from batteries, which also illuminate the night with our spotlights.

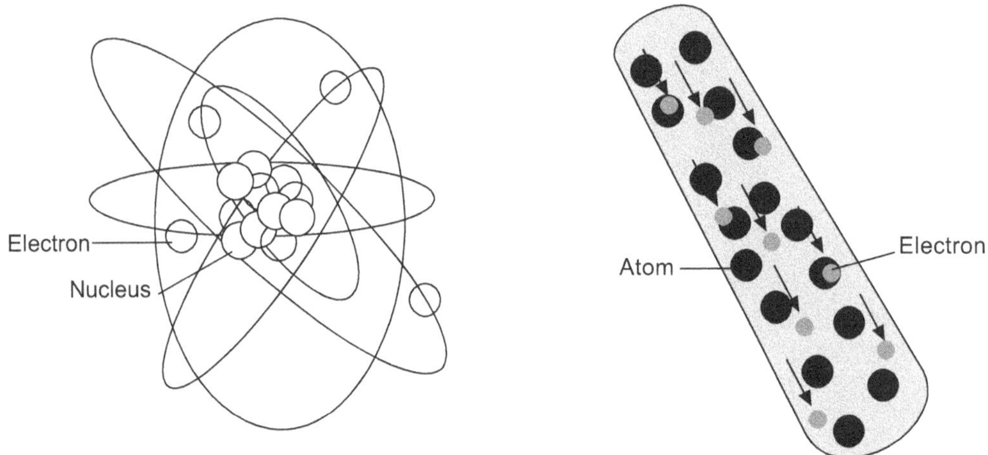

Fig. 4.1: Free movement of ions in metal

Atoms make up all matter, and smaller particles make up each atom. The proton, neutron, and electron are the three fundamental elements that make up an atom.

As the moon revolves around the earth, so do electrons around the nucleus, or centre, of atoms. Neutrons and protons comprise the nucleus.

Electrons are having the negative charge and protons are positively charged particle. Neutrons are neutral particles since they do not carry a positive or negative charge.

What relationship do positive and negative charges have to electricity, then?

Many techniques have been developed by scientists and engineers to produce lots of positive atoms and unbound negative electrons. Positive atoms have a strong attraction to the electrons because they need negative electrons to balance themselves.

The positive atoms are very attractive to the electrons since they also want to be a part of a balanced atom. In order to achieve equilibrium, the positive draws the negative.

The attraction between two particles becomes stronger the more positive atoms or negative electrons one has. We refer to the whole attraction, which includes both positive and negative charged units, as "charge."

4.2   WHICH WAY DOES THE "ELECTRICITY" REALLY FLOW?

In 1996, William Beaty wrote, "Electronics teachers and textbook authors are frequently criticised for imparting a "mistake" to their students." Instructors spread the myth that electric current is a flow of positive particles moving one way whereas, in reality, it's a flow of negative electrons.

As a matter of fact, the critics are mistaken and the professors are right. The chiders believe that electrons, which are always negatively charged, are the fundamental building blocks of "electricity." Because of this mistake, some individuals also mistakenly believe that electric currents are always a flow of negative particles.

Electric currents can actually flow with positive particles under certain conditions. Particles flow negatively in other circumstances. Moreover, they can occasionally flow in opposite directions at the same moment, both positive and negative. Depending on the conductor type, the particles' actual flow direction will change.

4.3 ELECTRICITY IS MORE THAN JUST ELECTRONS

Electric Charge, which is frequently referred to as "Quantity of Electricity," is not formed of electrons, to be more precise. Positive and negative particles both make up the two types of charge. These particles are provided by the protons and electrons in conductor atoms in the regular realm of electronics. Muons, positrons, antiprotons, and other particles may also be studied by physicists, but only protons and electrons are considered to be the "electricity" in typical electrical systems.

At this point, everyone will correctly point out to me that while electrons can move freely within wires, protons cannot. Certainly, this is true—but only in the case of metals, specifically those metals that are solid and not liquid. Positively charged atoms are the building blocks of metals and are submerged in a sea of mobile electrons. Although the protons within the positive atoms of copper do not move forward when an electric current is generated, the "electron sea" does.

Positive atoms participate in the electric current in many other things besides solid metals, which are not the only conductors. These different non-electron conductors are not particularly novel. They are as close to us as they possible can be, all around us.

4.4 NON-ELECTRON CHARGE-FLOW

For instance, if you stuck your fingers into a television's anode/fly back area, you would get a potentially fatal electric shock. There was undoubtedly a sizable current running through your body during the terrible event. But there was absolutely no electron movement through your body. Positive and negative charged atoms make up every single one of the electric charges in a human body. These atoms were the ones that moved as an electric current when you were being electrocuted.

The flow of the electric current included positive sodium and potassium atoms, negative chlorine atoms, and a large number of other more intricate positive and negative molcculcs. Positive atoms flowed in one direction together with the negative atoms that were concurrently flowing in the opposite direction during the electric current. Think of the flows as a collection of minute, moving dots, each of which is flowing in one of two directions.

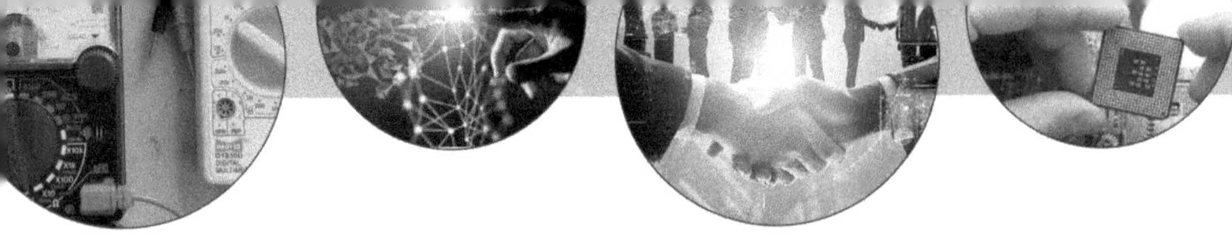

Without any dots colliding, the masses of tiny dots pass through one another. The negative atoms behave like electrons which drag an entire atom along with them, while without a single dot colliding with another, the masses of tiny dots flow through one another. Positive atoms behave like a proton with a full atom attached and negative atoms behave like electrons that drag an entire atom behind them.

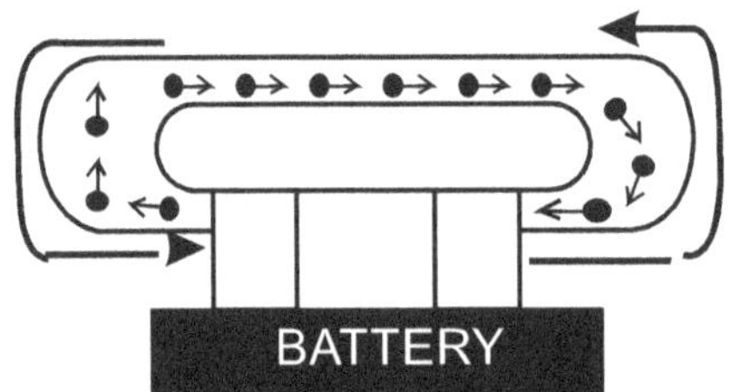

What way did the electric current actually travel in this scenario? Do we disregard the positive particles and follow the negative ones instead? or the opposite? here is an easy solution, but first...

Another illustration of a non-electron or "ionic" conductor is a battery. When a light bulb is connected to a battery, a complete circuit is created, and the flow of charge passes both through the filament of the light bulb and the interior of the battery. The electrolyte in batteries is particularly conductive.

The amperes of torch current appear as a flow of both positive and negative atoms deep inside the battery, within the moist chemicals between the plates. The battery is experiencing a strong electric charge flow, but there are no individual electrons moving through the battery at all.

What does the current actually do while it is flowing between the battery's two plates? not from left to right, but simultaneously in both directions. A little over half of the charge flow is made up of positive atoms, and the other half is made up of negative atoms moving backward.

The only true particle flow that occurs in the metal wires outside of the battery is from negative to positive. The charge-flow, however, simultaneously travels in two opposing directions inside the wet electrolyte of the battery. (And since we used hoses filled with salt water instead of wires, the current would flow in both directions.)

4.5 TWO-WAY CURRENTS ARE COMMON

This type of flow of positive and negative charges can be observed in many other places. Electric charges within conductors are a combination of moveable positive and negative particles in the following devices and materials. Both types of particles are moving in opposite directions past one another during an electric current.

Batteries and human bodies both include two-way positive and negative electric currents.

4.6   TWO-WAY POS/NEG ELECTRIC CURRENTS CAN EXIST

Every living thing

Liquid mercury and solder, Ion-based smoke detectors, Electroplating tanks, Electrophoresis gels in research, Air cleaners, smoke precipitations, The earth, the ocean, the sky (ionosphere), electrolytic capacitors, aluminum smelters,

Beams of particles

The atmospheric "sky current," which is a vertical flow of gas; Gas discharge, which includes:Neon signs, the Earth's aurora, lightning, corona discharges, electric sparks, fluorescent tubes, sodium and mercury arc streetlights, arc welders, Geigers counter tubes, thyratron tubes, and mercury vapour rectifiers are only a few examples of the phenomena that are mentioned.

Once more, may I ask you what the Actual direction of electrical current is?

We cannot resolve the issue by trivializing it or by acting as though two-way currents are only relevant to something foreign or outside of regular life. We cannot believe that a current in a wire is "real," but that a current in human flesh is not.

Students' confusion for 200 years

We can easily create electrical instruments, or "amp metres," that measure the Conventional Electric Current in terms of the magnetism that the charge-flow creates... or by the voltage drop that appears across a resistor, or by the temperature rise being created in a calibrated piece of resistance wire, once we start ignoring the speed and direction of the charges.

Charge-flow exists, but "AMPERES" do not.

No matter the particle polarities or fluxes, these three different types of metres will concur that a "current" is a "current." We can then utilise these metres anywhere. They will almost always provide us with all the information we could possibly need regarding charged particle fluxes in any circuit. In a complex physics experiment, an amp-meter might not be appropriate.

Designing electron beams inside vacuum tubes won't produce the desired results. It only monitors the simple current as it is typically defined, hence it cannot detect real current. Yet, for more than 99% of electricity and electronics, the particle orientation is irrelevant, and an ammeter just indicates the purportedly "real" current while concealing the actual particle fluxes.

Assuming that "electric currents" are always made up of positive particles, we characterise any negative currents as positive particles moving backwards rather than negatively.

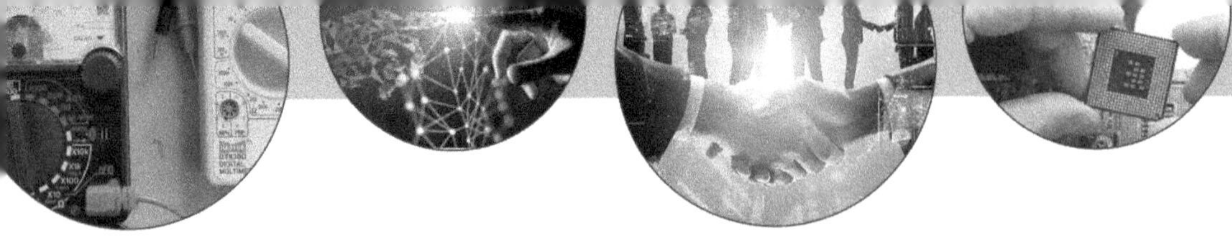

4.7   VOLTAGE AND CURRENT

If the circuit is complete, current will flow even while voltage tries to make it not. Voltage is occasionally referred to as the "push" or "force" of the electricity; while this isn't quite accurate, it might help you visualise what is happening. While current can move even in the absence of voltage, voltage cannot exist without current.

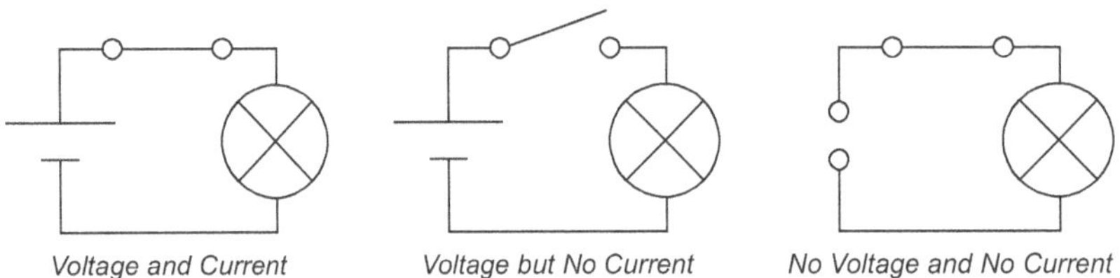

Fig. 4.2 : Voltage is the Cause, Current is the Effect

4.7.1   Voltage (V)

In electronics, we frequently refer to voltage at a spot as the voltage difference between that point and another. Voltage is the difference between two points. Voltage is a gauge of the amount of energy a charge. The correct word for voltage is potential difference, or p.d. for short, however this phrase is rarely used in electronics. Technically speaking, voltage is the "energy per unit charge.

"The battery supplies voltage (or power supply).

Components consume voltage, while cables do not.

We refer to a component's voltage across it.

Volts, or V, are used to express voltage.

A voltmeter that is connected in parallel is used to measure voltage.

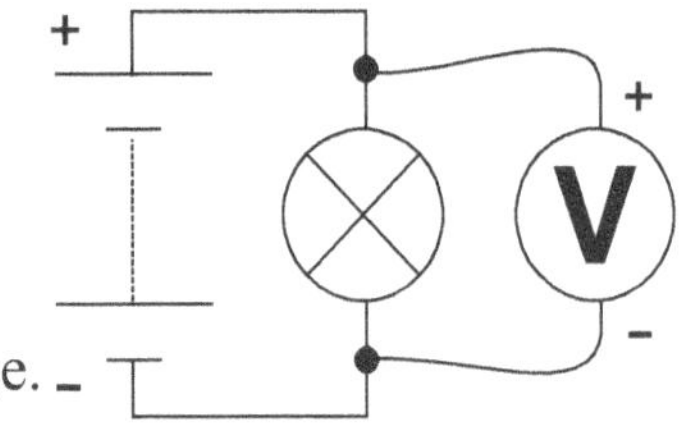

It should be noted that
    a) All components connected in series experience different voltages.
    b) All components connected in parallel have the same voltage.

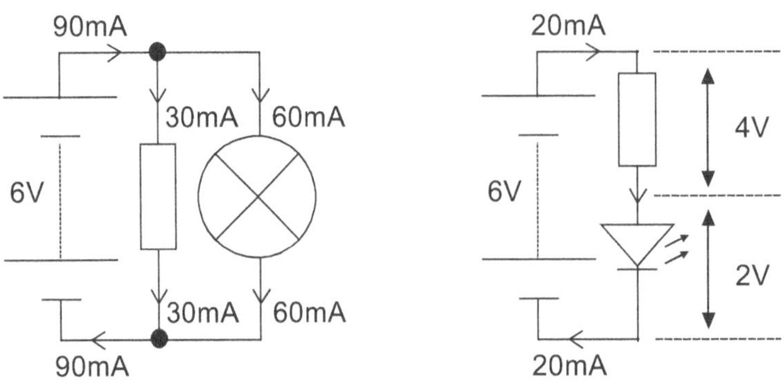

Fig. 4.3 : Voltage, Current in Series and Parallel

4.7.2 Current (I)

A component can only have input and output flow at the same rate, which is called current.

Amounts of current are expressed in amps (amperes).

A series-connected ammeter is used to measure current.As illustrated in the diagram, to connect in series, you must open the circuit and place the ammeter across the opening.Equations that deal with current utilise the sign I. Although 1A (1 amp) is a significant amount of current for electronics, mA (milliamps) are frequently utilised. M (milli) denotes a thousandth:

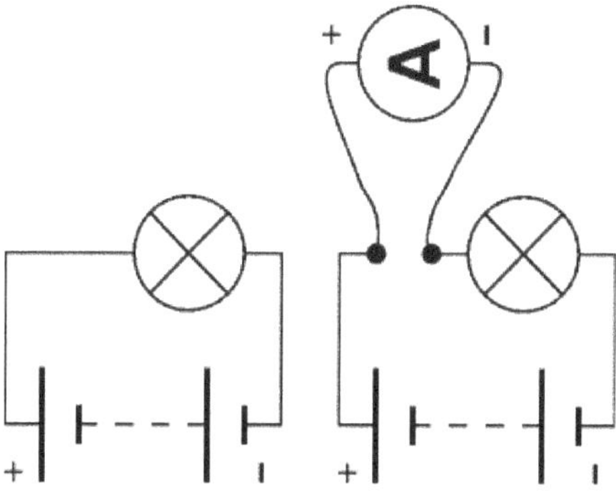

Fig. 4.4 : Connecting an ammeter in series

$$1mA = 0.001A, \text{ or } 1000mA = 1A$$

It should be noted that

   a) All components connected in series experience the same currents.

   b) All components connected in parallel experience distinct currents.

4.8 DIFFERENT TYPE OF ELECTRIC CURRENT

Depending on the flow direction, there are two different types of current:

(i) AC (ii) DC

Direct current is referred to as DC and alternating current as AC. When referring to voltages and electrical signals that are not currents, AC and DC are also utilised! A 12V AC power source, for instance, has an alternating voltage (which will make an alternating current flow). Electrical signals, which typically refer to voltages, are currents or voltages that carry information. Any voltage or current in a circuit may be referred to by this phrase.

4.8.1 Alternating Current (AC)

Electric current that alternates between flowing one way and the other is known as alternating current (AC). An AC voltage alternates between positive (+) and negative (-) all the time (-). The AC's frequency, which is defined as the number of forwards and backwards motions per second, is measured in hertz (Hz).cycle rates in reverse. For powering, an AC supply is appropriate.

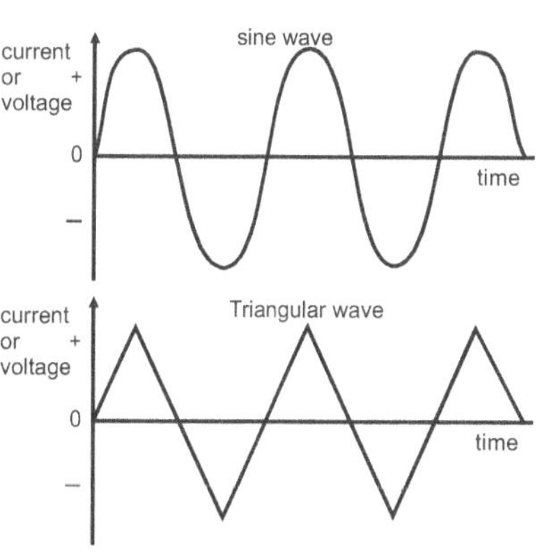

Fig. 4.5 : AC from a power supply

4.8.2 Direct Current (DC)

Direct Current (DC) may rise or decrease, but it always flows in the same direction.

A DC voltage may rise and fall, but it is always positive (or always negative).

Electronic circuits often require either a smooth DC supply with a little amount of variation termed ripple or a steady DC supply that is constant at one value.

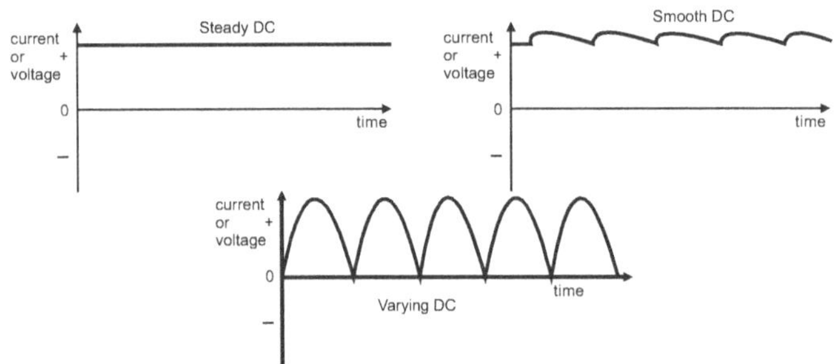

Fig. 4.6 : Waveforms for Steady and Varying Direct Current

Cells, batries, and controlled power sources all generate steady DC, which is ideal for electronic circuits. Different DC Supply wave forms are as follows in Fig. 4.6:

For electronic circuits, **steady DC** from a battery or controlled power source is optimal.
Certain gadgets can use **smooth DC** from a power source that has been smoothed out.
This is unsuitable for electronics since it **varies DC** from a power source without smoothing.

Power supplies have a transformer inside of them that changes the mains AC supply into a secure low voltage AC. A bridge rectifier is then used to convert the AC to DC, but the output is unstable DC, making it unsuitable for electronic circuits.

Certain power sources have a capacitor built in to deliver steady DC, ideal for less-sensitive electronic devices. Any DC supply will power lights, heaters, and motors.

4.9 THE ELECTRICAL SIGNAL'S PROPERTIES

Electrical signals, which typically refer to voltages, are currents or voltages that carry information. Any voltage or current in a circuit may be referred to by this phrase.

Figure 4.7's voltage-time graph demonstrates different electrical signal characteristics. In addition to the qualities highlighted on the graph, frequency, or the quantity of cycles per second, is also included.

Although a sine wave is depicted in the diagram, same characteristics apply to any signal with a consistent shape.

The signal's greatest voltage is its **amplitude**.
Volts (V) are used to measure it.
Amplitude is often referred to as **peak voltage.**
Peak voltage is doubled by **peak-peak voltage** (amplitude). Peak-peak voltage is typically measured when analysing an oscilloscope trace.

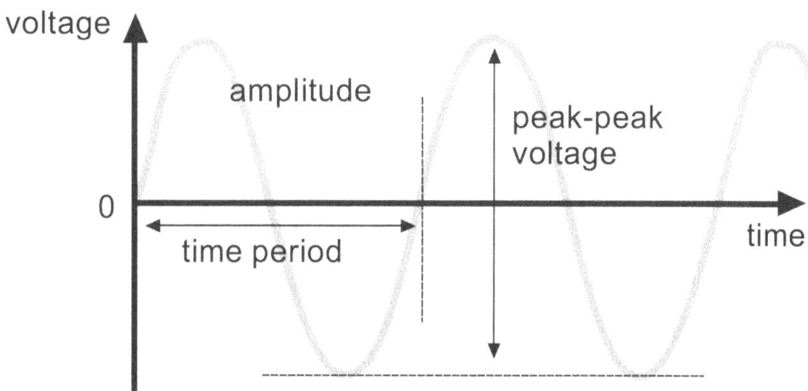

Fig. 4.7 : Wave properties

The signal's cycle time is measured in terms of **time period**.

It is measured in seconds (s), although since time intervals are frequently brief, milliseconds (ms) and microseconds (μs) are also frequently employed. 1ms = 0.001 s, and 1 μs = 0.000001 s.

The quantity of cycles per second is known as **frequency.**

Although frequencies are typically high, they are measured in **hertz (Hz),** kilohertz (kHz), megahertz. 1kHz = 1000Hz or 10^3 Hz and 1MHz = 1000000Hz or 10^6 Hz.

$$\textbf{Frequency=} \frac{1}{\text{Time Period}} \text{ and } \textbf{Time Period=} \frac{1}{\textbf{Frequency}}$$

Due to the **50Hz frequency** of UK mains power, its time period is 1/50, or 0.02 seconds, or **20 milliseconds.**

4.9.1 Root Mean Square (RMS) Values

From zero to the positive peak, the negative peak, and back to zero, AC voltage fluctuates continually. This is not a good indicator of its true effect because it is obvious that the majority of the time it is lower than the peak voltage.

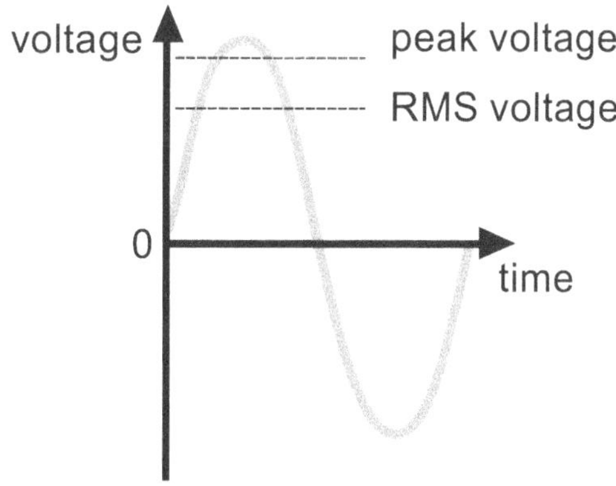

Fig. 4.8 Peak Voltage and RMS

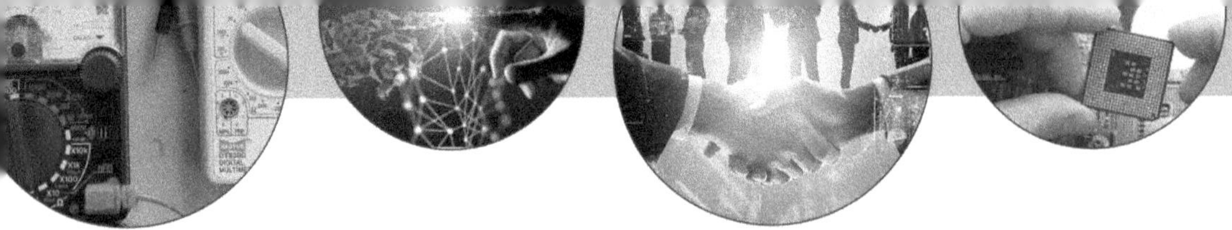

The root mean square voltage (VRMS), which is 0.7 of the peak voltage (Vpeak), is used instead:

$$V_{RMS} = 0.7 \times V_{peak} \quad \text{and} \quad V_{peak} = 1.4 \times V_{RMS}$$

These equations also apply to current.

The 0.7 and 1.4 have different values for other forms, therefore they only apply to sine waves, which are the most prevalent type of AC.

The effective value of a fluctuating voltage or current is known as the RMS value. The same effect is produced by the corresponding steady DC (constant) value.

If, for instance, the lamp in the circuit shown in Fig. 4.9. It is lit first by AC (by moving the 2 way switch to the left) and its brightness noted, then if 0.3 A DC produces the same brightness, the r.m.s. value of the AC is 0.3A. A lamp that is made to operate at 0.3A DC current will likewise operate at r.m.s. AC current value 0.3A. It can be shown that for a sine wave AC.

Please keep in mind that while it could be helpful to think of the RMS value as a sort of average, it is not the average. As the positive and negative components perfectly cancel out, the average voltage (or current) of an AC signal is zero.

4.9.2   RMS and AC Metre ?

AC voltmeters and ammeters display the voltage or current's RMS value. When linked to variable DC, DC metres will also display the RMS value as long as the variation is rapid; if the variation is slower than 10Hz, the metre reading will fluctuate.

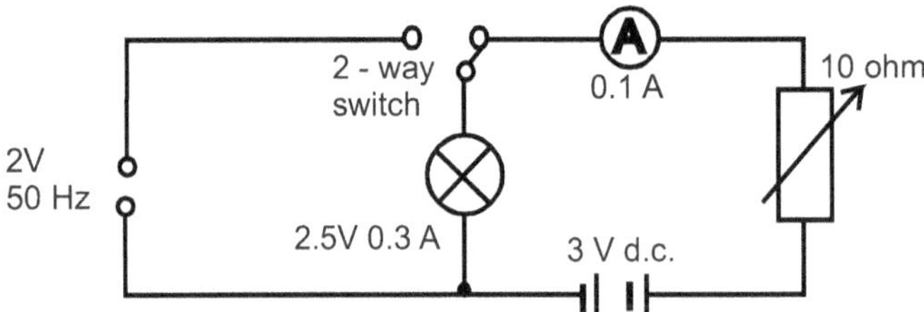

Fig. 4.9 : Root Mean Value of Lamp

r.m.s. value = peak value/2 = 0.7 peak value the peak value is much higher and is given by peak value = r.m.s. value/0.7
If the average value of a sine wave over half a cycles is required, it can be found from
RMS value equals 1.1 times the average value

4.10   UNIT OF CURRENT

Electric current is measured in unit of ampere (A) by an ammeter.

The current is 1 ampere (A), when 1 coulomb of charge passes through each point in a circuit in one second.

1 ampere = 1 coulomb/second (1 A = q/s)
In general if q coulomb as in t seconds, the current (I) in amperes is given by :
I = q/t OR q = It

Milliampere and micro ampere are the smallest units of current.

1 Milliampere (mA)	= 1/1000 A	= $1/10^3$ A	= 10^{-3} A
1 Microampere (mA)	= 1/1000000 A	= $1/10^6$ A	= 10^{-6} A
1 Nanoampere (nA)	= 1/1000000000 A	= $1/10^9$ A	= 10^{-9} A
1 Picoampere (pA)	= 1/1000000000000 A	= $1/10^{12}$ A	= 10^{-12} A

4.10.1 Symbols & Circuit Diagrams

Currents require complete conducting paths (circuits). Various symbols used in circuit diagrams are shown in Fig. 4.10

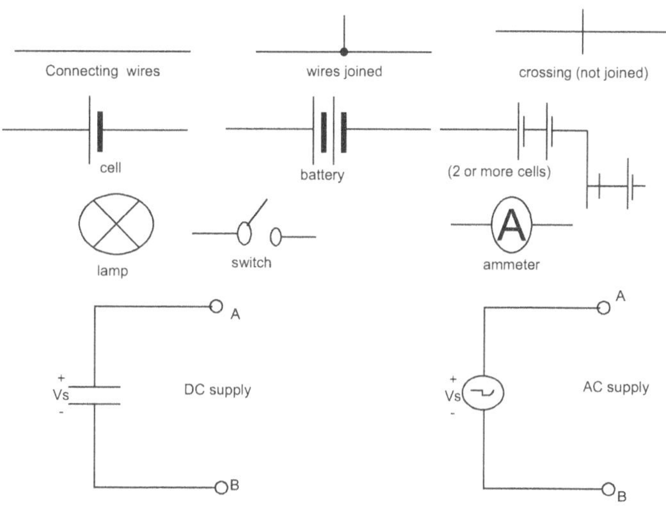

Fig. 4.10: Various Symbol Used in Circuit Design

4.11 ELECTROMOTIVE FORCE (EMF)

The electromotive force (e.m.f.) of a battery is defined to be 1 volt if it gives one joule of electrical energy to each coulomb of charge passing through it.

1 Volt = 1 joule/coulomb (1 V = 1J/C)

A 2 V battery gives 2 J of energy to each coulomb passing through it.

The e.m.f. of a battery can be measured by connecting a voltmeter across it as shown in Fig. 4.11

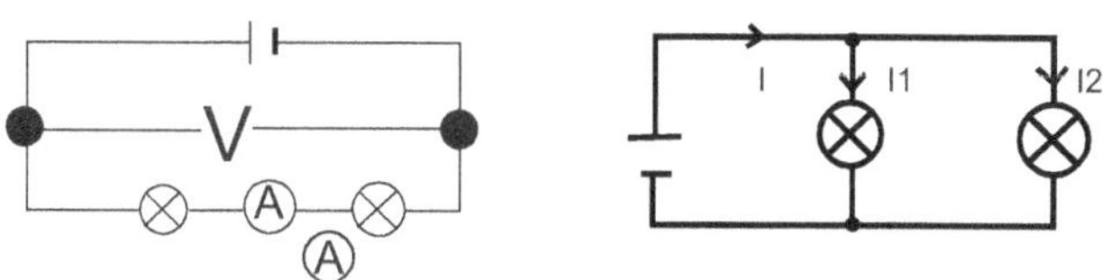

Fig. 4.11 Measurement of E.M.F. Using Voltmeter

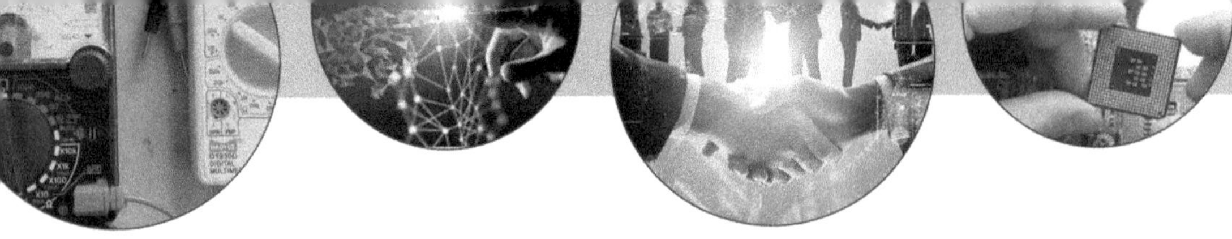

4.11.1 Electric Potential

The capacity of charged body to do work is called electric potential (V).

V = Work done/Charge = w/q

4.11.2 Potential Difference (P.D.)

Potential difference describes the variance in potential between two charged bodies. The potential difference (p.d.) across a device (e.g. a lamp) in a circuit is 1 volt if it changes 1 joule of electrical energy into other forms of energy (e.g. heat and light) when 1 coulomb passes through it.

$$V = w/q$$

Moreover, p.d. is measured by connecting a voltmeter across the apparatus as illustrated in Fig. 4.11 as both e.m.fs and p.ds. are measured in volts, they are commonly referred to as voltages.

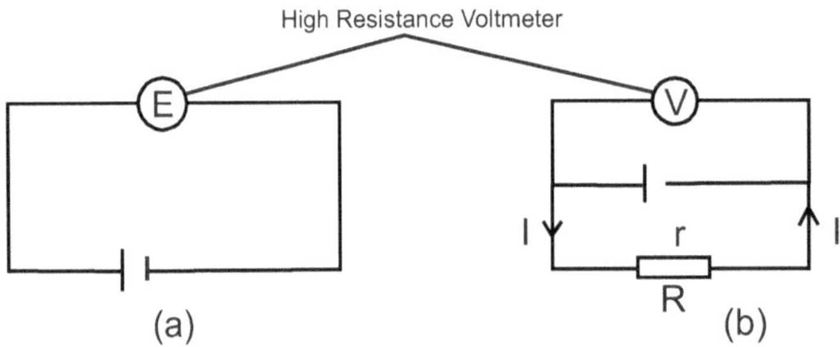

Fig 4.12 Measurement of Potential Difference (p.d.) Using volts.

4.11.3 Potential at a point

Although it is usually the p.d. between two points in a circuit we have to consider, there are times when it is a help to deal with the potential at a point.

When a point is selected in an electronic circuit to be at 0 volts, it is referred to as ground, common, or "earth" if it is connected to earth (e.g. via the earthing pin on a 3-pin plug). Fig.4.13 a, b lists the signs that were employed.

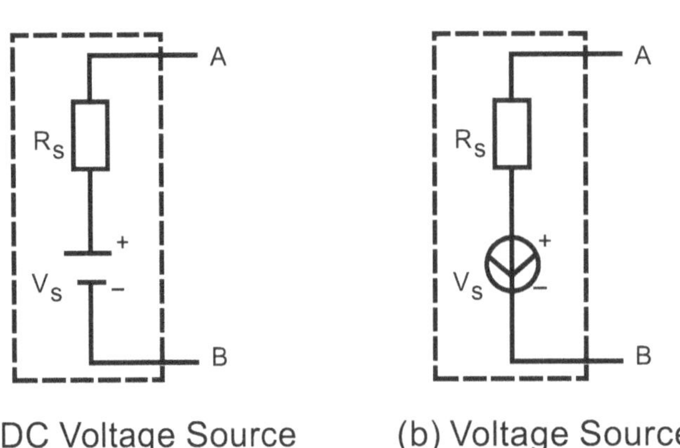

Fig. 4.13 : Potential at a Point

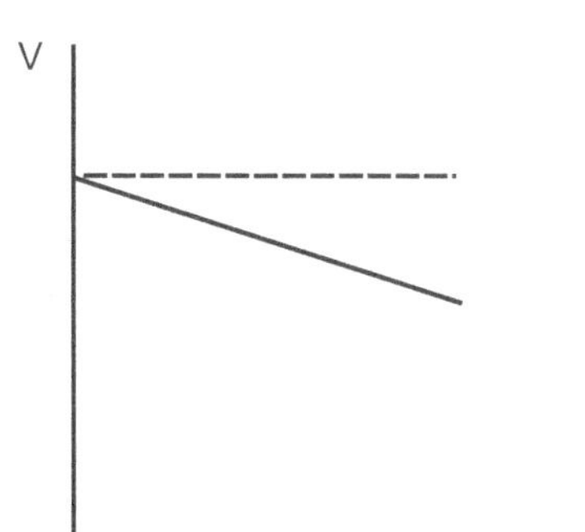

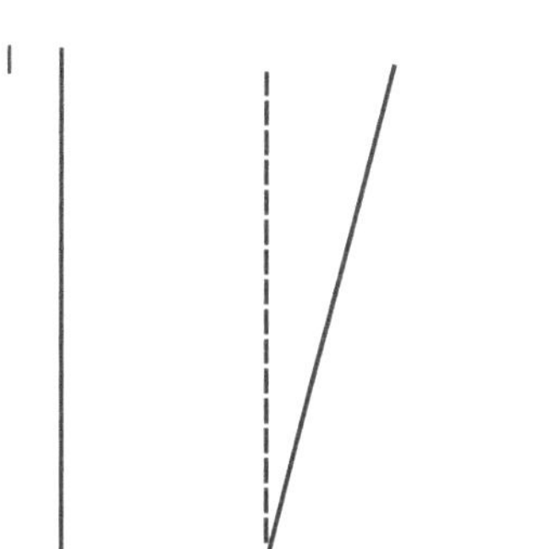

4.12 OHM'S LAW

German physicist Georg, discoverer of Ohm's law, the basic law of current flow. According to Ohm's law, the total resistance of a circuit, as shown in fig. 4.14, and the amount of current flowing in it are both inversely related to the electromotive force impressed on the circuit.

The law is usually expressed by the formula $I = V / R$, where I is the current in amperes, R is the resistance in ohms, while V or (E) is the electromotive force in volts. Ohm's law applies to all electric circuits for both direct current (DC) and alternating current (AC), but, in order to analyse complicated circuits and AC circuits that include contain inductances and capacitances, additional concepts must be applied.

Now take a look to a simple circuit:

Ex1 6 V is applied across a 10Ω resistor, what is the current?
　　　　$V = 6\,V,\ R = 10\,\Omega,\ I = ?,$
　　　　$I = V/R$
　　　　Current, $I = 6V/10\Omega = 0.6\,A$

Ex2 A 2.5 kΩ resistor passes a current of 0.2 A,
　　　what is the voltage across it?
　　　　　$V = ?, I = 0.2\,A, R = 2.5\,k\Omega = 2500\,\Omega$
　　　　　$V = I \times R$
　　　　　$V = 0.2 \times 2500 = 500\,V$

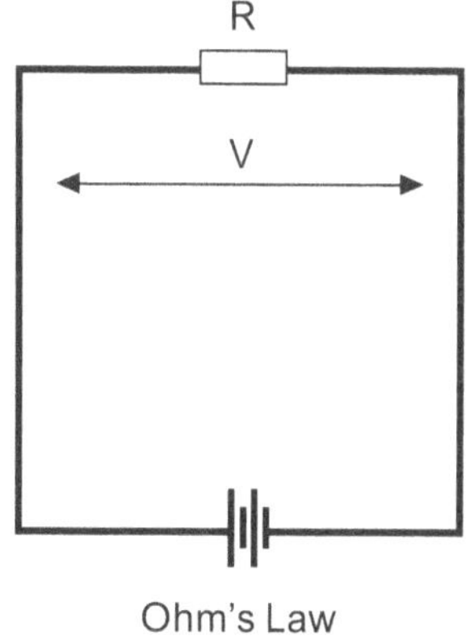

Ohm's Law

fig.4.14 : Ohm's Law

Ex3 A lamp connected to a 10 V battery passes
　　　a current of 40 mA, what is the lamp's resistance?
　　　　　$V = 10\,V, I = 40\,mA = 40 \times 10^{-3}, R = ?$
　　　　　$R = V/I$
　　　　　Resistance, $R = 10/40 \times 10^{-3} = 0.25\,k\Omega = 250\,\Omega$

German physicist G.S. Ohm made the connection between voltage, current, and resistance in a DC circuit. The current through a conductor is precisely proportional to the potential difference across it if the temperature is constant. This relationship is known as Ohm's law. This can be represented asOhmic or linear conductors obey Ohm's law and graph of I against V, called a characteristic curve, is a straight line through the origin (Fig. 4.15)

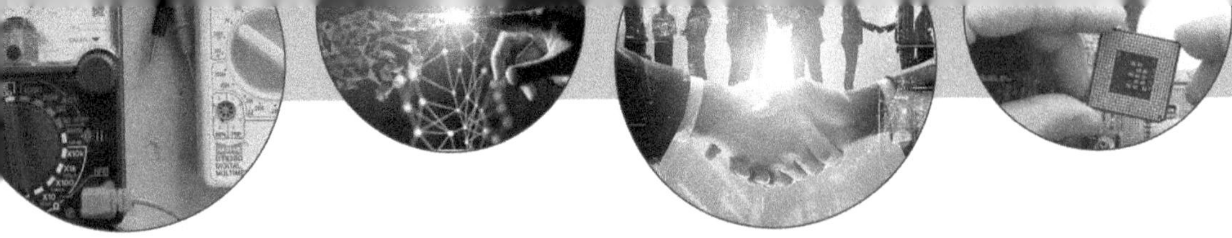

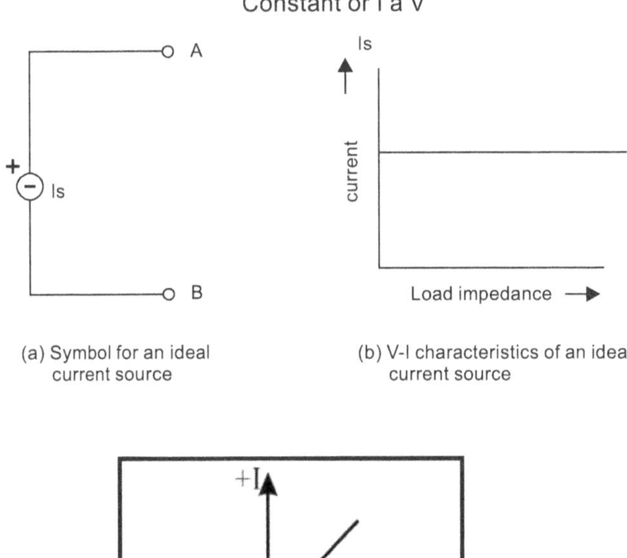

Fig. 4.15 : Ohm's Law and V-I characteristic Curve

4.13 ELECTRIC POWER

Energy use or production rate is referred to as power:

$$\text{Power} = \frac{\text{Energy}}{\text{Time}}$$

Power in watts (W)
Energy in joules (J)
Time in seconds (s)

The power of a device is 1 watt (1W) if it changes energy at the rate of 1 joule per second (1J/s) from one form to another. Every electronic component has a maximum power rating which should not be exceeded. For example a 1/2 W resistor can safely change 1/2 J of electrical energy per second into heat without damage.

A lamp like a torch bulb only needs around 1W, whereas an LED consumes approximately 40mW and a bleeper uses about 100mW. Power is typically measured in milli watts (mW), where 1mW = 0.001W, because electronics is primarily concerned with small amounts of power. V =IR, then

$$P = VI = IR \times I \ \textbf{(ohm's Law I=V/R)},$$

If R = 20 ohm and I = 1A

$$P = 1 \times 20 \times 1 = 20 \text{ W}$$

but if I = 2A, then $P = IR \times I = 2 \times 20 \times 2 = 80 \text{ W}$

Thus we see that by Doubling I quadruple P

To find the maximum safe current which can be passed through, for instance, a 5.0kΩ 0.5W resistor, we have

$$R = 5.0k\Omega = 5.0X1000\Omega \text{ and } P = 0.5 \text{ W}$$

P=VI =IR x I (ohm's Law I=V/R) OR $I^2 = P/R$

$$I^2 = 0.5/5000 = 5/5x10^4$$
$$= 1x10^{-4}$$
$$I = 10^{-2} A = 0.1 \text{ mA.}$$

4.13.1 Power Transfer

Internal Resistance

Power supplies have some resistance in themselves, called internal or source resistance. It causes the energy loss which occurs inside a battery(i.e. the 'lost' volts) when a current is driven through a load in a circuit. The greater the current the greater is the loss and the smaller is the terminal potential difference of the battery. Before we can calculate the maximum power supply deliver to a load of resistance R we need to know the e.m.f. E of the supply and its internal resistance.

In Fig. 4.16 A the high resistance voltmeter measure E (since the cell is on open circuit) and in Fig.4.16 A the high resistance voltmeter measure E (since the cell is on open circuit) and in Fig. 4.16 B it measures the terminal p.d. (V) which drives I through R1 i.e.since the current all around the circuit is I.

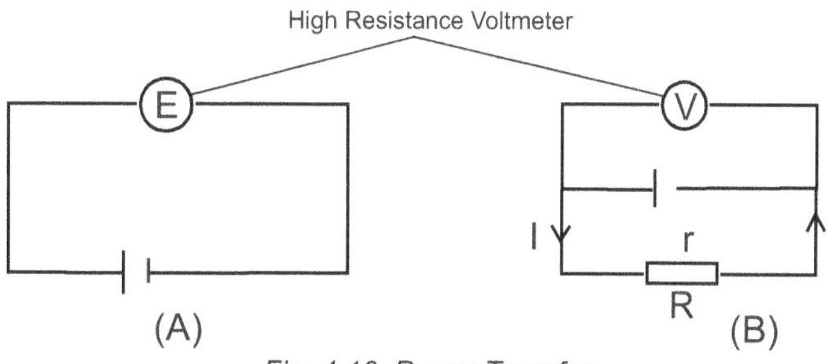

Fig. 4.16 Power Transfer

$$V = IR$$

$$\text{Voltage loss} = v = E - V = Ir$$

since the current all around the circuit is I., But E=V+v , E = IR + Ir = I (R+r)

I = E/(R+r) , P = I2R = E2R/ (R+r)2

4.13.2 Source of electric power

The basic purpose of source is to supply power to load. A source is, therefore, connected to the load as shown in Fig. 4.17 The source may supply either direct or alternating current.

Some DC source are battery, DC generator and rectification type DC supply. Similarly, AC sources are alternators, oscillators or signal generators.

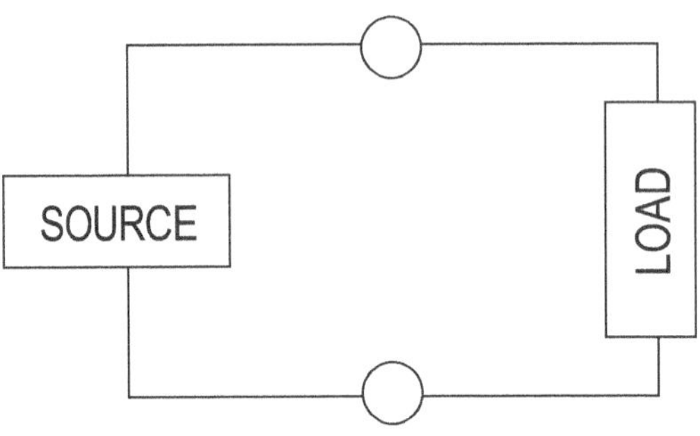

Fig. 4.17 Transfer of Energy from Source to Load

4.13.2.1 Batteries

The battery is the most common DC voltage source. It consists of a series or parallel combination of two or more similar cells. The basic generator of electrical energy is a cell. There are primary and secondary types of cells. While the primary cell cannot be recharged, the secondary cell can.

4.13.2.2 Generators

Compared to a battery, a DC generator is very different. It has a rotating shaft. When this shaft is rotated at a specified speed by some external agency, a voltage of rated value appears across its terminals. A generator is capable of giving higher voltage and power than a battery.

4.13.2.3 Alternators

The alternator is quite similar, both in construction and mode of operation, to the DC generator. The alternators are used in most electric power stations.

4.13.2.4 Oscillators or Signal Generators

An oscillator is the equipment which supplies AC voltage. It is used as a signal to test the working of different electronic circuits (such as an amplifier).

4.14 VOLTAGE DROP & INTERNAL IMPEDANCE OF A SOURCE

All electrical energy sources have some internal resistance or impedance. The source does not operate in a perfect manner because of this intrinsic impedance. When a voltage source supplies power to a load, its terminal voltage drops (voltage available at its terminal). When not attached to anything, a torch cell has a 2.0 V voltage across its two electrodes. However, when connected to bulb, its voltage becomes less than 2.0V.

In the equivalent circuit of Fig. 4.18, the bulb is replaced by load resistor R_L (of say, 0.8Ω) and the cell is replaced by a constant voltage source of 2.0V in series with the internal resistance R_S (of say, 0.2Ω). The total resistance in the circuit is now $0.2+0.80=1.0\Omega$. The current in the circuit is because the net voltage of 2.0 V, which sends current into the circuit.

Equation: $I = V/R = 2.0/1.0 = 2.0A$

The terminal voltage across the terminals AB) of the cell is same as the voltage across the load resistor. Therefore,R_L

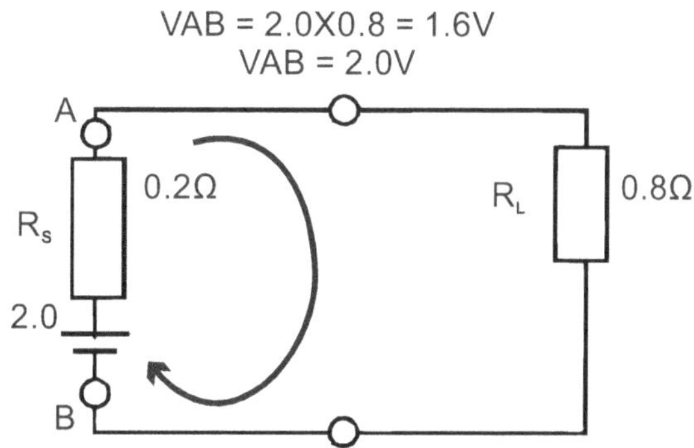

Fig. 4.18 : A cell of 1.5 V Connected to a Bulb.

The internal resistance causes a voltage drop that is Voltage drop = 2.0 - 1.6 = 0.4V

The internal resistance of a source may be due to one or more of the following reasons :

(i) The resistance of the electrolyte between the electrodes, in case of a cell.

(ii) The resistance of the armature winding in case of an alternator or a DC generator.

(iii) The active device's output impedance, such as that of a transistor or vacuum tube, is a

case of an oscillator (or signal generator), and rectification type DC supply.

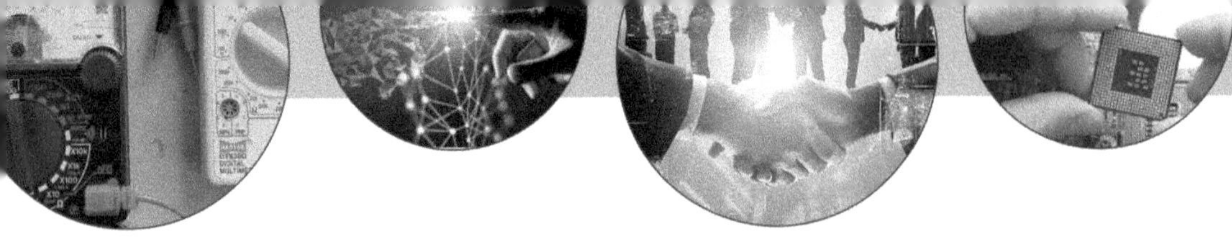

4.15 KIRCHHOFF'S LAWS

Two additional rules are used to determine the current flowing in each branch of a circuit if there are multiple interconnected branches. Kirchhoff's rules of networks are the names given to these laws, which the German physicist Gustav Robert Kirchhoff found.

4.15.1 Kirchhoffs First Law

The algebraic sum of currents incoming and outgoing any point in a circuit must equal The first of Kirchhoff's laws asserts that the sum of the currents equals the sum of the flows at each junction in a circuit when a constant current is flowing rates from the point to the currents flowing away from that point. Now take a look to the equation in each point of the circuit::

$$A - I_2 = I_1 + I_3$$
$$B - I_6 = I_2 + I_4$$
$$C - I_4 = I_3 + I_5$$
$$D - I_5 = I_1 + I_4$$

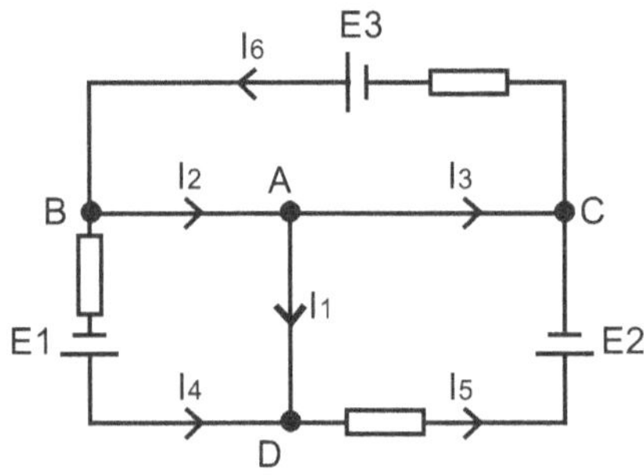

Fig. 4.19 : Kirchoff's First Law

4.15.2 Kirchhoffs Second Law

Any closed path's surrounding voltages add up to zero algebraically. According to the second law, the net sum of the electromotive forces that are countered will equal the net amount of the electromotive forces that are applied along any closed channel from any point in a network back to the starting point. The byproducts of the currents passing through them and the resistance they faced. It is merely a development of Ohm's law. Now take a look to the equation for each path :

Starting B point :

$$E_3 = R_2 x I_6$$

Starting D point :

$$E_2 = R_1 x I_5$$

Starting A point:

$$E_1 = R_3 x I_4$$

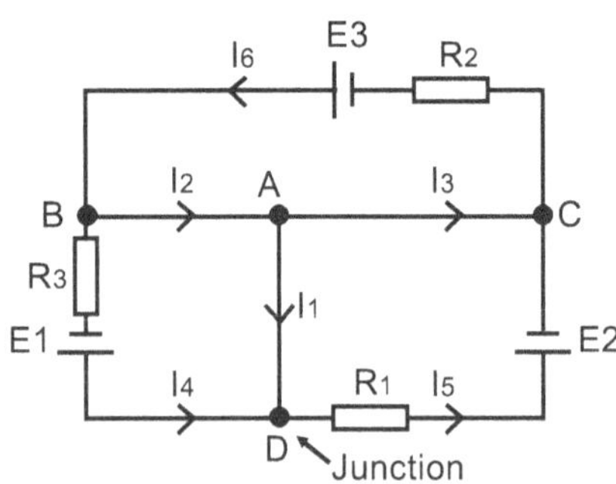

Fig. 4.20 : Kirchoff's Second Law

According to the first of them, all closed paths or loops in the network have zero total voltage at any one time. According to the second, the sum of the currents arriving at any given point in time at any node, or intersection of pathways, in a network is equal to the sum of the currents leaving at that same point in time.

4.16 CONSEQUENCES OF KVL AND KCL

Voltage Dividers

There is a voltage drop across each element of a circuit when two elements are connected in series, but the current through both elements must be the same. The resistances determine how the voltage is divided at each link in the chain. A voltage divider is a straightforward circuit that consists of two (or more) resistors connected in series with a source.

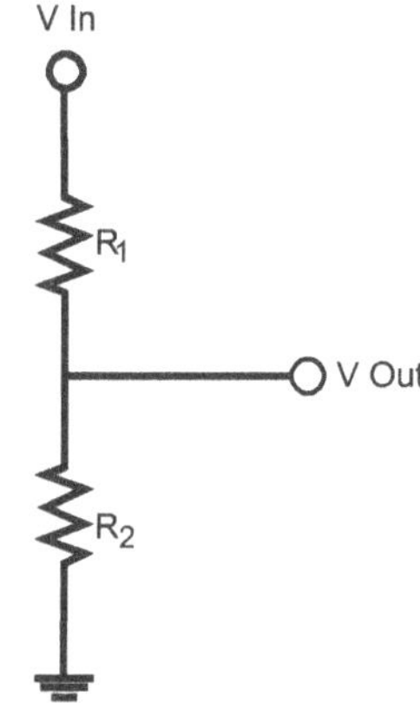

Fig. 4.21: Voltage Divider circuit

Consider the circuit in Figure 4.21 According to KVL the voltage Vin is dropped across resistors R_1 and R_2 Ohm's Law applies if a current I travels through the two series resistors. So

$$i = \frac{V\ In}{R_2 + R_1}$$

$$V_{out} = i\ R_2$$

Therefore

$$V_{out} = \frac{V_{IN}\ R_2}{R_1 + R_2}$$

If voltage across R1 is VR1 then

$$V_{R1} = \frac{V_{IN}\ R_1}{R_1 + R_2}$$

Typically, the voltage dropped across one of the n series resistors, say Ri, is

$$V_{R1} = \frac{V_{IN}\ R_i}{R_{eq}}$$

$$R_{eq} = R_1 + \ldots + R_n$$

4.16.1 Voltage dividers as references

Voltage dividers can undoubtedly be used as references. For example, if you need 5.5 volts from a 11.0 volt battery, connect two equal-valued resistors in series and place the reference between the second resistor and ground. Although there are undoubtedly further issues, the first one is current draw and how the source impedance affects the circuit.

If the source impedance is, for example, 50Ω, connecting two 100Ω resistors is clearly not a good idea. If the battery is rated at, say, 200 milli ampere hours, the current draw would be fairly high at 0.044 mA. Moreover, the reference voltage is less comfortable with that source impedance, making the loading more inconvenient. $V_{out} = 4.4V = IR_2$

$$3.6V = \frac{11}{250}$$

So clearly increasing the order of the resistor to at least 1 k ohm is the way to go to reduce the current draw and the effect of loading. The other problem with these voltage divider references is that the reference cannot be loaded if we put a 100 ohm resistor in parallel with a 10 kohm resistor, when the voltage divider is made of two 10 kohm resistors, then the resistance of the reference resistor becomes somewhere near 100 Ω. This clearly means a terrible reference.

If a 10 MΩ resistor is used for the reference resistor will still be some where around 10 kΩ but still probably less. The impact of tolerances is another issue; for example, 10 kΩ resistor's resistance can vary by about 500Ω if the resistors are rated at 5% tolerance. This results in even more inaccurate references of this kind.

4.16.2 Current Distributors in Parallel

The voltage across two elements when hey are connected in parallel The stream must be the same, yet it is divided by the resistance. A current divider is a straightforward electrical circuit that consists of two (or more) parallel resistors and a source.

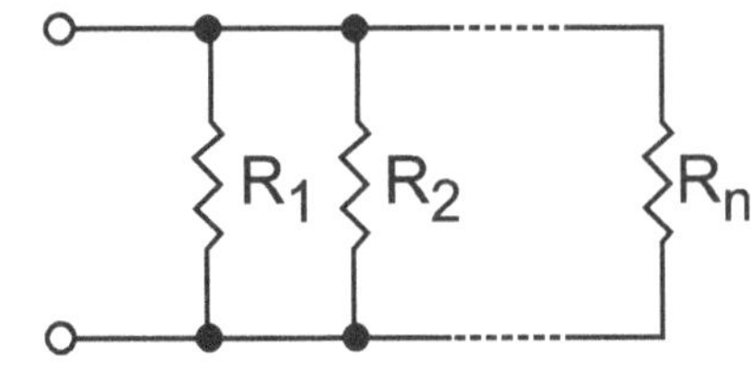

Fig. 4.22 : Parallel registers

If a voltage V appears across the resistors in Figure 4.22 with only R_1 and R_2 for the moment then the current flowing in the circuit, before the division, i is according to Ohms Law.

$$i = \frac{V}{R_{eq}} \quad \ldots\ldots\ldots\ldots(1)$$

A parallel set of resistors is used with the equivalent resistance.

$$i = \frac{V(R_1+R_2)}{R_1 R_2} \quad \ldots\ldots\ldots\ldots(2)$$

According to Ohms Law, the current via R1 equals

$$i_1 = \frac{V}{R_1}$$

Dividing equation (2) by (1)

$$i_1 = \frac{iR_2}{R_2+R_1}$$

Similarly

$$i_2 = \frac{iR_1}{R_2+R_1}$$

The current ix with n Resistors is typically

$$i_x = \frac{iR_1\,R_2\cdots R_n}{(R_2\cdots R_n + \cdots + R_1\cdots R_{n-1})R_x}$$

Or possibly more simply

$$\frac{i_x}{i} = \frac{R_{eq}}{R_x}$$

Where

$$R_{eq} = \frac{R_1\cdots R_n}{R_2\cdots R_n + \cdots + R_1\cdots R_{n-1}}$$

CHAPTER-5

Magnetism

Magnetism and Magnetic fields

5.1 MAGNETISM AND MAGNETIC FIELDS

Some uncommon stones known as lodestones were first identified to be naturally magnetised by the ancient Greeks and Chinese. When allowed to swing freely while suspended by a piece of string, these stones were found to magically attract tiny pieces of iron and constantly point in the same direction. Thessaly, Greece's Magnesia, is where the name of the place originates. From the 1600s to the present, numerous scientists have made significant contributions to our knowledge of magnets and their properties.

Magnetism is a phenomenon that appears to have nothing to do with electricity because of how compasses and the earth's magnetic field interact, as well as because of refrigerator magnets and magnets on children's toys, we are all familiar with magnetism. Similar language to that used to describe electric forces is used to describe magnetic forces:

1. There are two types of magnetic poles, conventionally called North and South

2. Different poles repel whereas similar poles attract.
   Magnetism, however, differs from electricity in one crucial way:

3. There are no magnetic monopoles; instead, magnetic poles always appear in North-South pairs, unlike electric charges.

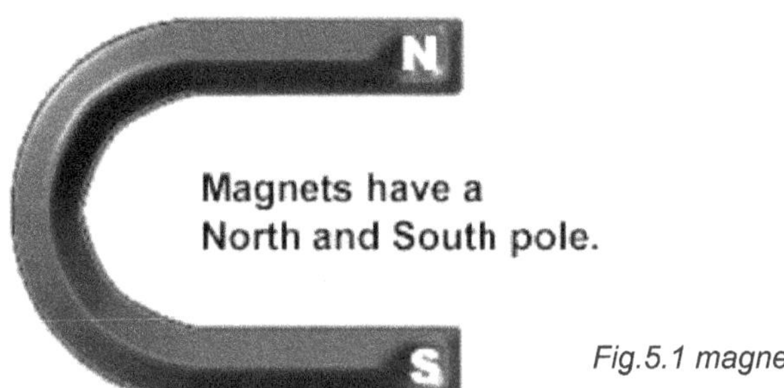

Fig.5.1 magnet

For discussing the behaviour of magnetic forces, it is convenient to use the analogy of an electric charge, as in the case of electric charges.

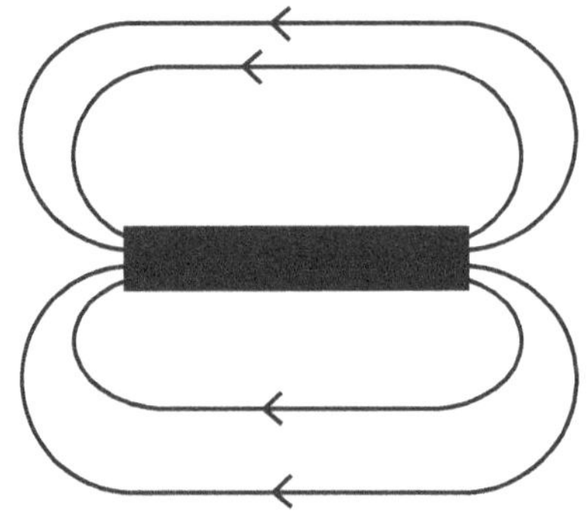

Fig. 5.2 Lines of force surround the magnet

Figure 5.1 shows the magnetic field lines for a bar magnet.

These lines can be interpreted by students as showing in which direction a compass needle will point if it is positioned at that location.Tesla units are used to express the strength of magnetic fields (T). The earth's magnetic field is on the order of 0.0001 T, which is actually a quite strong field.

5.1.1 Ten Facts about Magnets:

1. The north and south poles point in opposite directions.
2. Polar opposites attract and opposite poles repel.
3. Only magnetic materials are drawn to magnetic forces.
4. Magnetic forces operate remotely.
5. Temporary magnets behave like permanent magnets while they are magnetised.
6. A coil of wire becomes a magnet when an electric current passes through it.
7. Adding iron to a coil that carries current strengthens the magnetic.
8. An electric current is induced in a conductor by a shifting magnetic field.
9. A charged particle does not experience a magnetic force when it moves parallel to a magnetic field, but when it moves perpendicular to the field, it experiences a force that is opposed to the motion's direction and the field.
10. A magnetic field that is perpendicular to a wire that is carrying current is subject to a force that is perpendicular to the wire and the field.

5.2 TYPES OF MAGNETS:

There are three main types of magnets:

> Permanent magnets
> Temporary magnets
> Electromagnets

5.2.1 Permanent Magnets

We are most familiar with permanent magnets, such as those that hang from our refrigerator doors. They are permanent in the sense that they retain some magnetic after being magnetised. As we shall see, many kinds of permanent magnets have various traits or features with regard to how readily they may be demagnetized, how powerful they can be, how their strength fluctuates with temperature, and so forth.

5.2.2 Temporary Magnets

Temporary magnets are those that behave like permanent magnets when they are surrounded by a powerful magnetic field but lose their magnetic properties when the magnetic field is removed. Paperclips, nails, and other objects made of soft iron are a few examples.

5.2.3 Electromagnets

An electromagnet is a coil of wire that is tightly wound into a helical shape and typically has an iron core. While current is running through the coil, it behaves like a permanent magnet. By altering the amount of current flowing through the wire and the direction of the current flow, the electromagnet's magnetic field may be made to have different strengths and polarities.

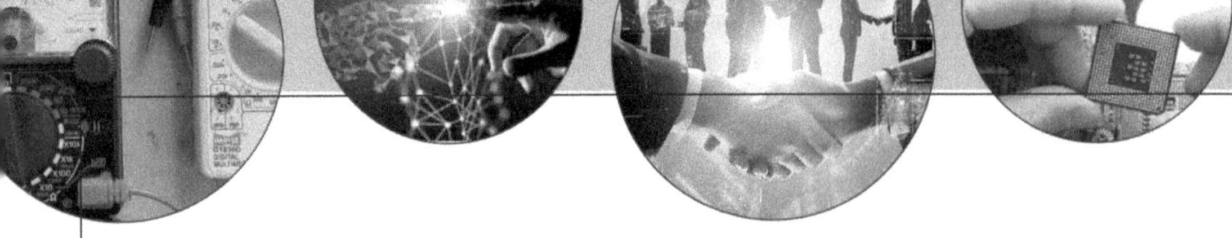

5.3 MAGNETIC FIELD SOURCES

Electric currents, which can be microscopic currents linked to electrons in atomic orbits or macroscopic currents in wires, generate magnetic fields. The Lorentz force law defines the magnetic field B in terms of the force acting on a moving charge. There are numerous practical uses for the magnetic field and charge interaction. Dipolar in nature, magnetic field sources have a north and south magnetic pole. The magnetic component of the Lorentz force law, Fmagnetic = qvB, indicates that the SI unit for magnetic field is the Tesla, which is made up of (Newtons per second)/. (Coulomb x meter). The Gauss is a unit of magnetic field size (1 Tesla = 10,000 Gauss).

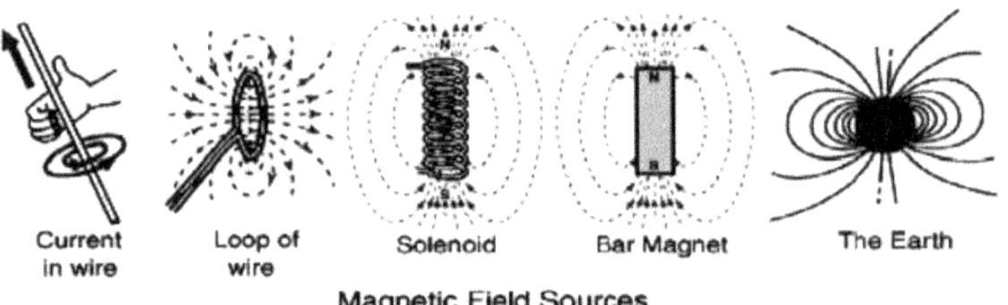

5.4 ELECTRICITY AND MAGNETISM

All electromagnetic fields are force fields that may produce an action at a distance and carry energy. Both wave and particle features can be found in these domains. A wire or coil carrying an electric current generates its own magnetic field. The needle of a compass will swing perpendicular to the wire. Why are magnetic fields different from electric fields? Actually both are usually all around us, since they are present wherever there is electricity. Both an electric and a magnetic field exist around power lines, appliances, light fixtures, and electric wiring; but electromagnetic fields are generated only when current is flowing through a wire. The voltage on a wire produces electric fields as shown in fig. 5.3 for example if you plug in a lamp or an appliance, even though it is not turned on, it will produce an electric field. When the light is lit or the appliance is running, both an electric and electromagnetic field will be present

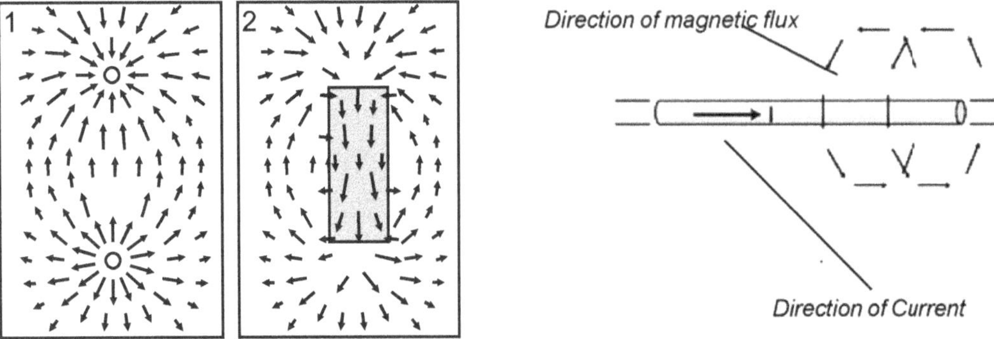

Fig. 5.3, 1-Electric fields, 2- Magnetic fields

5.5 SOME ELECTROMAGNETIC DEVICES

We use in our daily life many magnetic or electromagnetic devices as given below:

(1) Headphones (2) Stereo speakers (3) Computer speakers (4) Telephone receivers (5) Phone ringers (6) Microwave tubes (7) Doorbell ringer (8) Solenoid (9) Object holders made of refrigerator magnets Refrigerator door seal (10) Plug-in battery drains Audio tape recording and playback head, video tape recording and playback head, and floppy disc recording and reading head (12) Magnetic stripe for credit cards coil for TV deflection (13) TV degaussing coil (14) Computer monitor deflection coil Computer hard drive recording and reading head (15) Dishwasher water valve solenoid Shower curtain weights / attach to tub (16) Power supply transformers

5.6 RELATION BETWEEN CURRENT FLOW AND MAGNETIC FIELDS

This can be explained with the help of Right Hand Rule for magnetic field from flowing current, and for magnetic field in a coil.

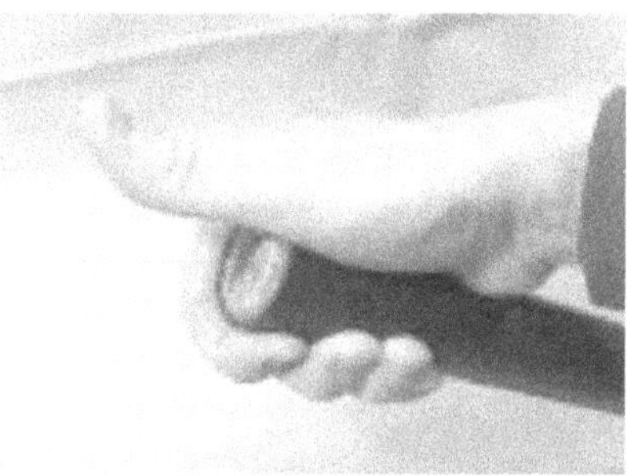

Fig. 5.4 Current and magnetic field

As seen in fig. 5.4, a magnetic field forms around a wire as current runs through it. Using your right hand, curl the fingers, and extend the thumb to help you envision this. Now, point your thumb in the direction that the wire's current is flowing (using conventional current, which has the current flow from a battery's + end to its - end). The magnetic field around the wire is oriented in the same direction that your fingers are curled. Your thumb and fingers would point in a clockwise direction to the magnetic field around the wire, respectively, if the current were flowing directly from this page towards you. **Where there is a current there will be magnetic field and vice versa.**

5.7 ELECTROMAGNETIC INDUCTION

To understand the concept of electromagnetic induction, the fig.5.5 shows an ammeter is connected in the circuit of a conducting loop. An electromotive force (emf) is induced in the loop when the bar magnet is pushed towards or away from it. Depending on how the magnet and loop are moving in relation to one another, the ammeter shows currents flowing in various directions. Take note that the current returns to zero when the magnet stops moving, as shown by the ammeter.

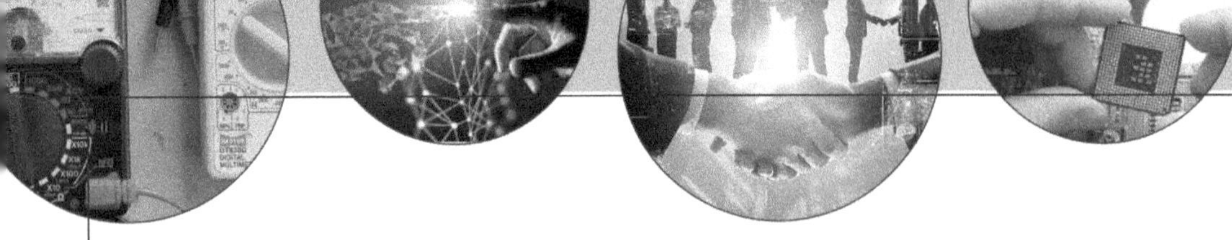

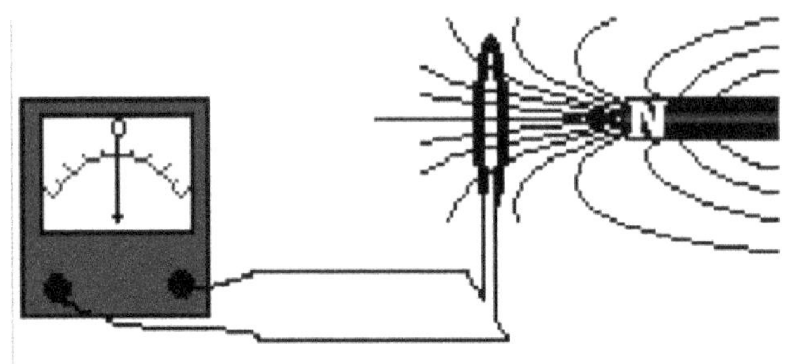

Fig. 5.5 Electromagnetic induction

This proves that a conductor's potential difference (voltage) is caused by a changing magnetic field. The term "electromagnetic induction" refers to this idea. Electric generators, also known as alternators, transformers, Tesla coils, microphones, and electric guitars all operate on the electromagnetic induction principle. You should also be aware that because the current in this conductor flows back and forth, it is alternating current. The conductor was lifted and then dropped in the magnetic field, which is what caused this. You should now have a better understanding of why our home's present alternates as a result.

5.8 TRANSFORMER

A "transformer" changes one voltage to another. This attribute is useful in many ways. It doesn't change power levels. 100 Watts will come out the opposite end of a transformer if you put 100 Watts into it. (approx 95%)

Two coils of wire coiled tightly around an iron or ferrite "core" are what make up a transformer. One coil (the "primary") receives power, which produces a magnetic field.

It should be noted that direct current (DC) cannot be used for this; instead, alternating current (AC) or pulsed DC must be used.

The transformer's ability to adjust voltage depends on how many times the wires are wrapped around the core (or "turns").

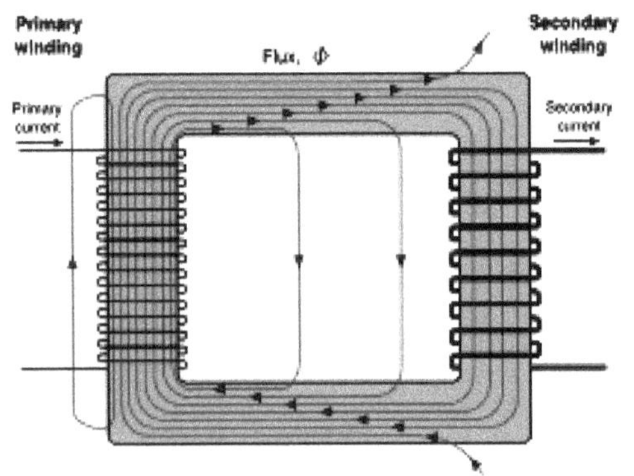

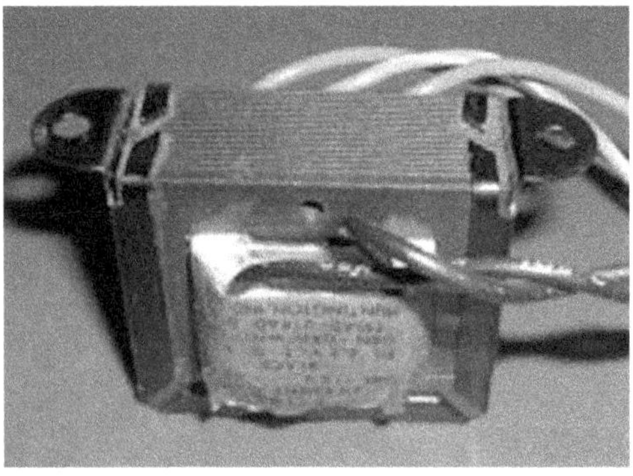

Fig.5.6 two-winding transformer & Laminated core transformer

If the primary has less turns than the secondary, you have a step-up transformer that increases the voltage. A step-down transformer lowers the voltage if the primary has more turns than the secondary.

The outgoing voltage will match the incoming voltage if the primary has the same number of turns as the secondary. An isolation transformer falls under this category.One big coil of wire can function as the primary and secondary in some rare circumstances. This is true for xenon strobe trigger transformers and variable auto-transformers.

5.9   HOW TRANSFORMER WORKS

The transformer is built on two fundamental ideas: first, that an electric current may generate a magnetic field (electromagnetism), and second, that a fluctuating magnetic field inside a coil of wire causes a voltage to be induced at the coil's ends (electromagnetic induction). As the secondary coil is encircled by the same magnetic field as the primary coil, a voltage is induced across the secondary when the primary coil's current is changed.

A simplified transformer design is shown in Figure 5.6 A current passing through the primary coil creates a magnetic field. The primary and secondary coils are wound around a core of exceptionally high magnetic permeability, such as iron, to ensure that the majority of the magnetic field lines created by the primary current are within the iron and pass through both the primary coil and the secondary coil.

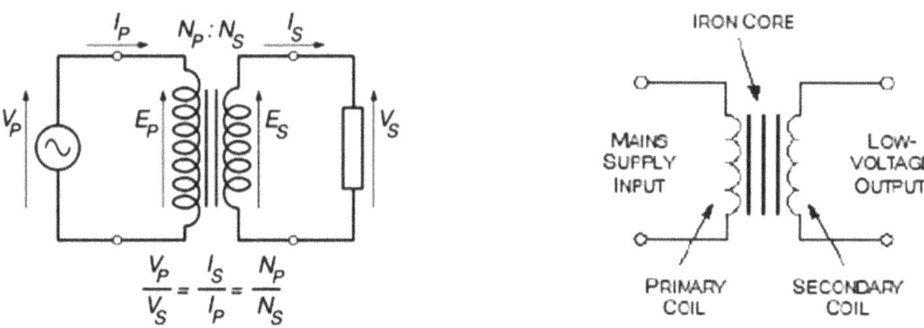

Fig.5.6 ideal transformer as a circuit element

Electrical power is transferred from the primary circuit to the secondary circuit if the secondary coil is connected to a load that permits current to flow.

Incorrectly referred to as "induced electromotive force," often known as "induced e.m.f.," is the induced voltage.

5.10   POWER OF TRANSFORMER

The transformer should operate at its maximum efficiency, converting all incoming energy from the main circuit to the magnetic field and subsequently to the secondary circuit. The incoming electric power must match the exiting power if this criterion is true. Ideal equation is given

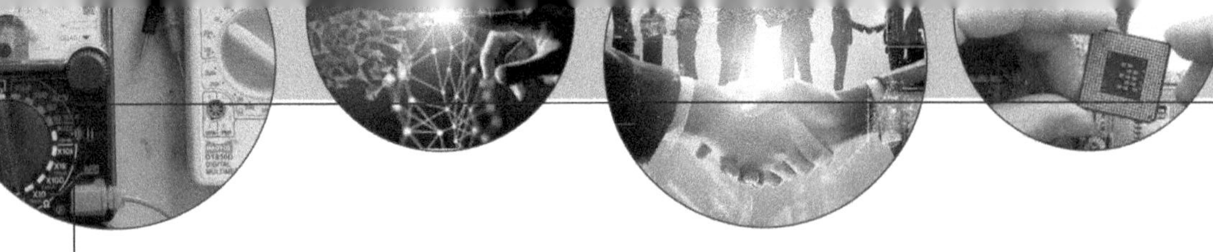

$$P_{incoming} = I_P V_P = P_{Outgoing} = I_S V_S$$

$$\frac{V_S}{V_P} = \frac{N_S}{N_P} = \frac{I_P}{I_S}$$

Where:

Vp= Voltage in Primary coil
Vs= Voltage in Secondary coil
Ip= Current in Primary coil
Is= Current in Secondary coil

Np=No. of turns in primary coil
Ns=No. of turns in Secondary coil
Ep=electromagnetic field in primary coil
Es=electromagnetic field in Secondary coil

To stop the flow of "eddy currents," the iron core is laminated. These currents are generated when the alternating magnetic field induces a minor voltage in the core, similar to the voltage created in the secondary coil. Eddy currents waste energy by unnecessarily heating the core, but they are significantly diminished by laminating the iron because this raises the core's electrical resistance without changing its magnetic characteristics. Compared to alternative techniques of adjusting voltage, transformers have two significant benefits:

1. They completely isolate the electrical input from the output, allowing for the safe reduction of the high mains supply voltage.

2. A transformer wastes almost no energy. They have a high efficiency (power out / power in) of 95% or more.

5.11 TYPES OF TRANSFORMERS

Transformers are typically employed for two things: power supplies and signal matching. Transformers are built with qualities that are appropriate for the use for which they are designed. The size of the windings or the connection between the primary and secondary windings may fluctuate depending on the construction. The purpose a transformer serves in a circuit, such as an isolation transformer, is another way to categorise different sorts of transformers.

5.11.1 Power Transformers

At high power levels, power transformers are utilised to convert between different voltages.

5.11.2 Step-up Transformers

A gadget that needs a high voltage power source can run on a lower voltage source thanks to a step-up transformer. A low voltage input is transformed by the transformer into a high current at a high voltage output.

For example: The CRT display tube of your computer monitor requires thousands of volts, but must run off of 110 VAC from the wall.

5.11.3 Step-down Transformers

A device that needs a low voltage power source can run from a greater voltage by using a step-down transformer. The transformer produces a low voltage at a high current by taking in a high voltage at a low current.

For Example:

• Your Mobile charger run on 5 DC, but you plug them into the 220 VAC line.

• Your doorbell doesn't need batteries. It operates on 12 volts of 220 volts AC.

5.11.4 Isolation Transformers

No voltage is raised or lowered by a isolation transformer ; whatever voltage enters is what exits. Without an isolation transformer, current cannot flow directly from one side to the other. Often, this acts as a safety feature to guard against electrocution.

5.11.5 Variable auto-Transformers

Variacs, also known as variable auto-transformers, can function as step-up or step-down transformers. You can select any output voltage you like with the large knob on top.

(5.11.6: Inverters

An inverter raises the voltage of a DC power supply from low to high. The most common type of inverter takes power from an automobile and cranks out 110 VAC to run appliances and power tools. Inventors are also used to operate fluorescent lamps from battery power. In reality, an inverter contains a transformer; it is not a transformer itself (and lots of other stuff).

5.11.7 Signal Transformers

Signal transformers are devices that take something in and change it into something else to come out. The altered object in this instance, however, contains some sort of information signal and the power levels are low.

5.12 ENERGY LOSSES IN TRANSFORMER

Transformer losses can be divided into two categories: those coming from the magnetic circuit, often known as iron loss, and those coming from the windings, commonly referred to as copper loss. Losses can also be stated as "no-load," "full-load," or at an intermediate loading, and they change with load current. The main cause of load losses is winding resistance, but hysteresis and eddy current losses account for more than 99% of no-load losses. Because of the potential for high no-load loss, even an idle transformer uses up electrical energy, spurring the creation of low-loss transformers.

5.12.1 Winding resistance

The conductors heat up resistively as a result of current flowing through the windings. Skin effect and proximity impact increase winding resistance and losses at higher frequencies.

5.12.2 Losses in hysteresis

Hysteresis in the core causes a tiny quantity of energy to be wasted each time the magnetic field is reversed. For a given core material, the loss is a function of the peak flux density to which it is exposed and is proportional to the frequency.

5.12.3 Eddy currents

A solid core constructed of a ferromagnetic material is an excellent conductor as well, and it likewise has a single short-circuited turn along its whole length. As a result, eddy currents go through the core in a plane normal to the flux, which causes the core material to heat up resistively. The inverse square of the material thickness and the square of the supply frequency are intricate functions that determine the eddy current loss.

5.12.4 Magnetostriction

The phenomenon known as magnetostriction occurs when the magnetic flux in a ferromagnetic material, such as the core, causes it to physically expand and contract slightly with each cycle of the magnetic field. This results in losses owing to frictional heating in vulnerable cores, which is what gives transformers their distinctive buzzing sound.

5.12.5 Mechanical losses

Together with magnetostriction, the fluctuating electromagnetic forces between the primary and secondary windings are also influenced by the alternating magnetic field. They cause neighbouring metals to vibrate, which increases the buzzing noise and uses a little amount of power.

5.12.6 Stray losses

Since the energy that powers its magnetic fields is returned to the source with each half-cycle, leakage inductance is by itself lossless. However, any leakage flux that collides with nearby conductive elements, like the support structure of the transformer, will result in eddy currents and heat.

CHAPTER-6

Semiconductors

6.1 SEMICONDUCTORS MATERIALS

They are the solids that are sandwiched between good insulators (rubber) and good conductors (copper). Materials like silicon, germanium, cadmium sulfide, and gallium arsenide are examples of semiconductors.

Elements from the periodic table with electrical characteristics that are intermediate between conductors (copper, aluminium) and insulators (glass,rubber). Electronic gadgets made of semiconductor materials are also described by the word. Digital audio players, mobile phones, and computers are a few examples. Although dozens of other materials are also used, silicon is the primary material used in the production of most semiconductors.

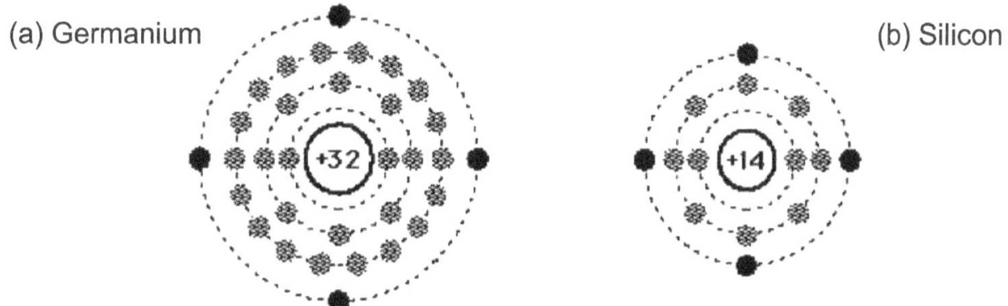

Fig. 6.1 : Atomic Structure of Germanium, Silicon

6.1.1 Structure of Semiconductor Materials

Figure 6.1 depicts the electrical structure of pure silicon or germanium, both of which have four valence electrons in their outermost orbits (b). We can see in Fig. 6.1(b) that the Si atom shares its four valence electrons with the following four Si atoms to form an octahedral structure. Nonetheless, germanium will have more free electrons and a better conductivity at a given temperature. Since silicon can operate at far greater temperatures than germanium, it is by far the more popular semiconductor for electronics.

6.1.2 Moderate capacity of Semiconductors

Semiconductors have a similar social impact despite being invisible. Solid-state electronics, however, is transforming technology at the same time as technology has undoubtedly altered society. Semi-conduction, which is a significant electrical feature of a relatively small group of elements and compounds, describes a state in which these substances are neither excellent electrical insulators nor excellent electrical conductors. Rather, they have a moderate capacity for electrical conductivity. Just think about a world without technology.

6.1.3 Reduction in Size with high speed

Radios, TVs, computers, video games, and subpar medical diagnosis tools wouldn't exist. Although vacuum tube technology could be used to create a variety of electronic devices, advances in semiconductor technology over the past 50 years have allowed for the creation of smaller, quicker, and more dependable electronic devices.

All the times you have interacted with technology devices. Our lives could continue to be made better by developments in the world of technology. Understanding and being able to engage in the domains of communication, computers, medicine, the basic sciences, and engineering are made possible by learning about electronic materials. Electronics are widely used in each of these fields.

6.2 VALANCE ELECTRONS

Valence electrons are the electrons that make up an atom's outermost shell; they control the nature of an atom's chemical processes and play a significant role in determining the electrical properties of solid matter. The band theory of solids depicts the electrical characteristics of matter in terms of how much energy is required to liberate a valence electron.

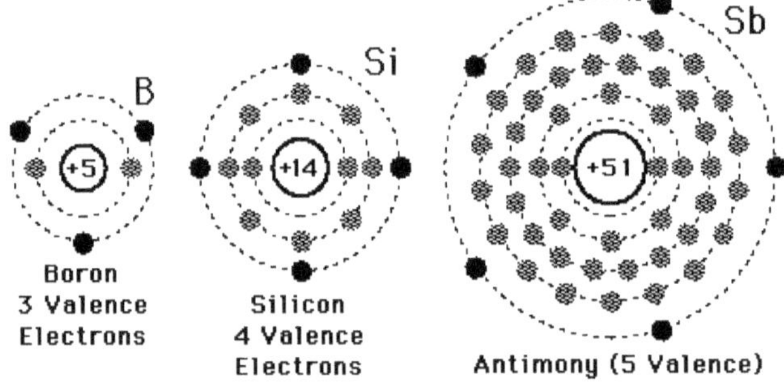

Fig. 6.2 : Valence Electrons

6.3 ATOMIC SILICON LATTICE

Silicon atoms can crystallise into a regular lattice and form covalent bonds.

The true crystal structure of silicon is a diamond lattice; the figure below is merely a simplified representation of that structure.

This crystal, known as an intrinsic semiconductor, can carry a very little current.

A silicon atom contains four electrons, which it can share in covalent connections with its neighbours. This is the key point.

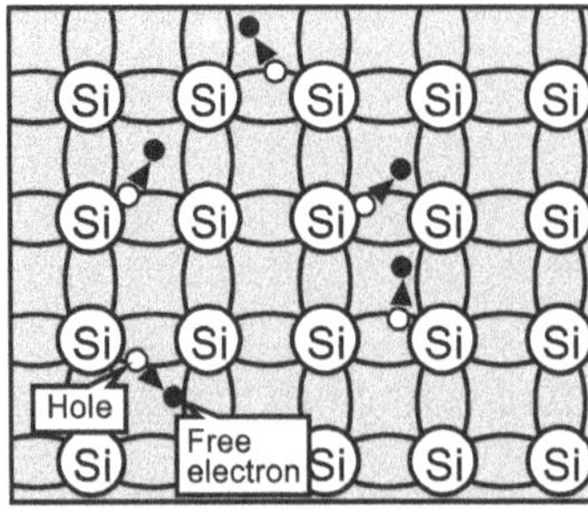

Fig. 6.3 : Silicon Lattice

6.4 INTRINSIC SEMICONDUCTORS

As seen in fig. 6.4, intrinsic semiconductors are a pure semiconductor devoid of any appreciable impurity species.

1. In intrinsic semiconductors, there are equal amounts of electrons and holes. These simple depictions do not adequately capture the nature of that sharing because each silicon atom will be influenced by more than four other silicon atoms, as can be shown by looking at the silicon unit cell equal.

2. Crystal flaws or thermal excitation can be the cause of intrinsic semiconductors' conductivity.

3. The number of holes in the valence band and the number of electrons in the conduction band are identical in an intrinsic semiconductor, as depicted in Fig. 6.3.

4. Both the electron and the hole can contribute to a tiny current flow if a voltage is provided. it is not employed commercially.

5. As may be seen in fig. 6.4, intrinsic semiconductors function similarly to insulators.

As a result, pure silicon and germanium are excellent insulators, being ideal at temperatures close to absolute zero, as shown in fig. 6.4.

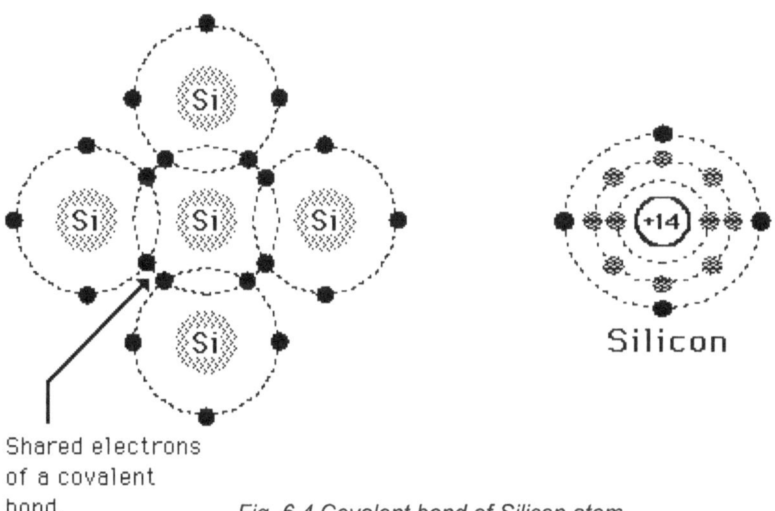

Fig. 6.4 Covalent bond of Silicon atom

6.5 EXTRINSIC SEMICONDUCTORS

Extrinsic semiconductors are those that have undergone impurity doping to alter the types and numbers of free charge carriers.

In semiconductor manufacturing, tiny quantities of boron, phosphorus, or arsenic are added to crystals to add extra electrons (donor impurity) or holes (acceptor impurity). also known as a "dopant."

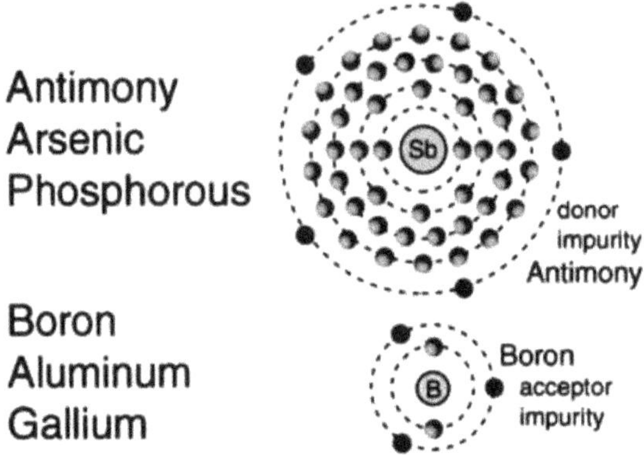

Fig. 6.5 Extrinsic Semiconductor

The normal crystal lattice of silicon or germanium is dramatically altered by the addition of a small quantity of foreign atoms, resulting in n-type and p-type semiconductors. according to Fig. 6.5

Pentavalent contaminants (Sb, As, P)

N-type semiconductors are created by the extra electrons added by impurity atoms with five valence electrons.

Trivalent contaminants (B, Al, Ga)

P-type semiconductors are created by impurity atoms having three valence electrons, or "holes," or electron deficiencies..

6.5.1   N-Type Semiconductor

When we combine pure Silicon (Si) with penta valent antimony (Sb), the bulk of the electrons form an n-type semiconductor. With five valence electrons, an antimony atom. The effects of adding one antimony atom to the silicon crystal's lattice are depicted in Fig. 6.6. Because it is held loosely, the fifth valence electron is free and can take part in conduction.Four of its valence electrons establish covalent bonds with four of its nearby silicon atoms.

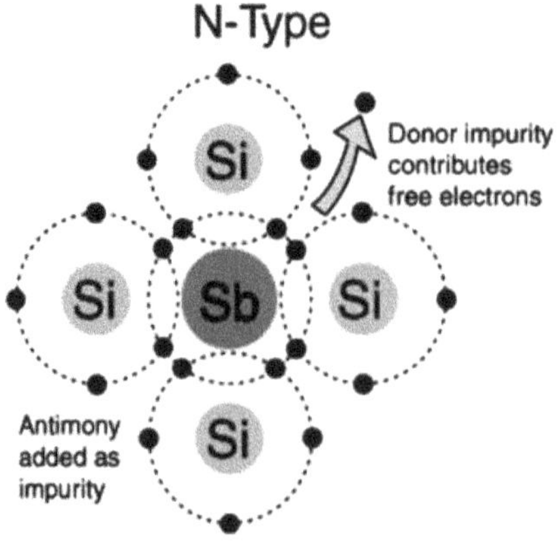

Fig. 6.6 N-type Semiconductor

The conductivity of the intrinsic semiconductor is considerably increased by the addition of pentavalent impurities like antimony, arsenic, or phosphorous.

The donor impurity, which is ionised and gains a positive charge, is imprisoned in the lattice and unable to move. Nonetheless, when an electric field is applied to the semiconductor and a current is flowing, it can still have an impact on the migration of electrons. Later on in the course, this will be covered in more detail. As the vast majority of donors will be ionised, it can often be assumed that the electron concentration in an n-type semiconductor at room temperature is just equal to the total concentration of all donors.

6.5.2 P-Type Semiconductor

When pure silicon (Si) and trivalent Boron (B) are combined, the resultant semiconductor is referred to as a "p-type semiconductor" because holes predominate over all other semiconductor characteristics. Three valence electrons make up the boron atom. The effects of adding one Boron atom to the silicon crystal's lattice are depicted in Fig. 6.7. Its four valence electrons are spare and can participate in conduction because three of them establish covalent bonds with three nearby silicon atoms. In P-type materials, holes thus make up the bulk of carriers while electrons make up the minority. A type IIb (blue diamond) that has boron (B) impurities is an illustration of a naturally occurring P-type semiconductor.

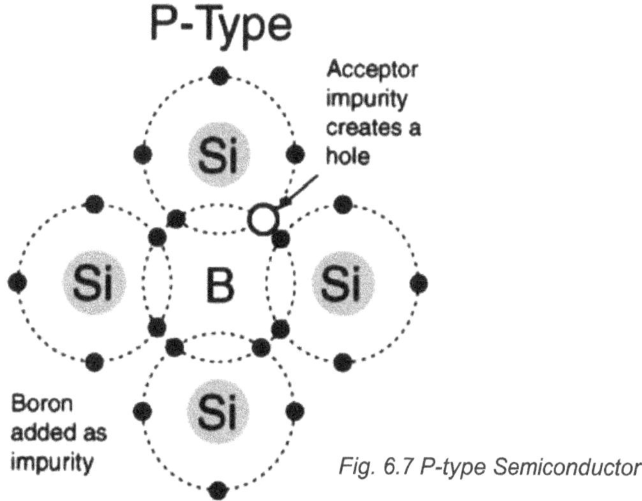

Fig. 6.7 P-type Semiconductor

6.6 PN-JUCTION

As shown in fig. 6.8, a p-n junction is made by combining two semiconductors, one of which is doped n-type and the other p-type. The holes prefer to diffuse from the P type material to the N type material, while the electrons from the N type material covertly diffuse from the p type material. This minority carrier diffusion eventually comes to an end because it has given rise to a space-charge zone, which prevents further charge diffusion and fixes the minority carriers in place. The PN junction diode's depletion area is referred to as this.

A built-in potential is produced by the space-charge area that the minority carriers produce.

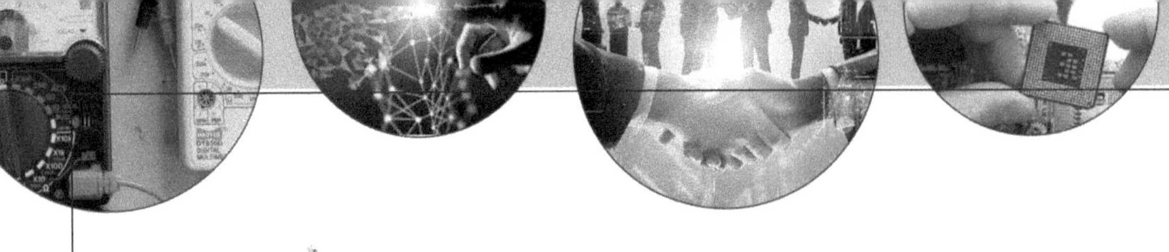

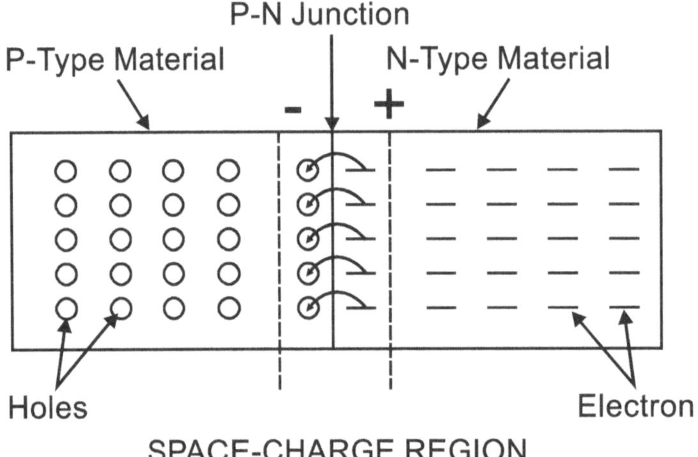

Fig. 6.8 Holes and Electrons Interaction at PN- Junction

We receive 0 V when we place a voltage metre over a PN junction. A voltage is obtained by passing a voltage metre across a patch of space charge. why is that?

We receive a voltage across the space-charge area as a result of the potential gradient driving the drift current, or flow of charge carriers, in the opposite direction of the diffusion current. The net current across the p-n junction is zero when equilibrium circumstances exist because the drift current and diffusion current are precisely balanced. The majority carriers are unable to conduct current across the junction due to the increase of ions. Due to the minority carriers using the junction, the current flow over the barrier is not nearly zero.

6.7 External Voltage and PN- Junction

Electron-hole pairs are produced when the crystal is exposed to an external energy source (such as light, heat, etc.). Minority current carriers are created by pairs of electron-holes. In both locations, there exist small current carriers: holes in the N material and electrons in the P material.

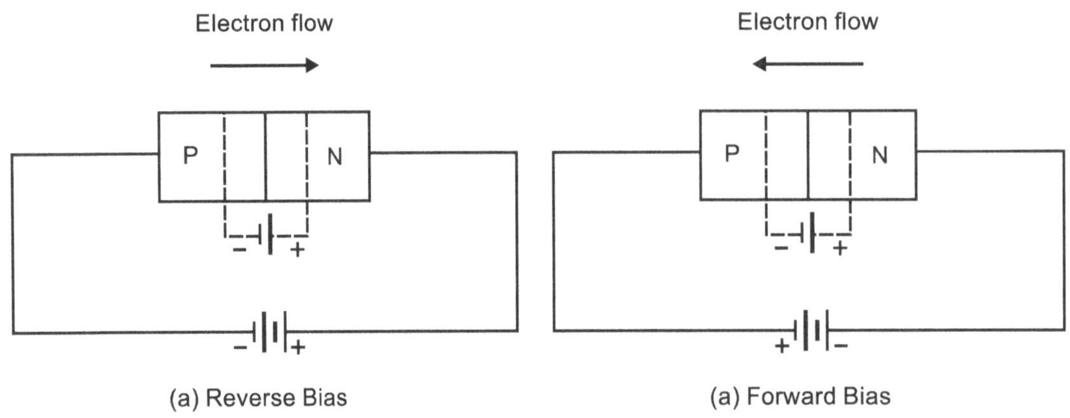

Fig. 6.9 Current Flow in PN-Junction

The negative terminal of the battery acts as a repellent for the electrons in the P-type material under reverse bias, as shown in fig. 6.9. An ion in the N-type material will be neutralised by the electron as it travels over the junction.

Similar to this, the positive battery terminal will cause the holes in the N-type material to reject one another and move away from the junction. In the P-type material, a negative ion will be neutralised when the hole reaches the junction.

As the involved holes and electrons originate from the electron-hole pairs that are produced in the crystal lattice structure and not from the addition of impurity atoms, this flow of minority carriers is known as MINORITY CURRENT FLOW.

As a result, when a p-n junction is reverse biased, majority carriers will not cause any current flow, while minority carriers will cause very little current to flow through the junction. This little current, though, might be disregarded at typical working temperatures.

The p-n junction diode's ability to provide minimal resistance to current flow when biased forward but maximal resistance when biased reverse is, in essence, the most crucial thing to keep in mind about it.

CHAPTER-7

PN-Junction diode

7.1 PN-JUnCTION DIODE

The earliest type of semiconductor device was a crystal detector, which was first used in wireless radios. In order to make this invention, German scientist Ferdinand Braun utilized a single metal wire, sometimes known as a "cat's whisker," to touch a semiconductor crystal. The final product, a "rectifying diode" (so termed due to the fact that it has two terminals), easily permits current to flow in one direction but limits flow in the other. Vacuum-tube diodes, however, had almost entirely replaced the tiny but much quirkier crystal detector by 1930.

7.2 WORLD WAR AND PN-JUNCTION DIODE

Despite their instability, crystals were more effective than vacuum-tube diodes at rectification of the high frequencies used by radar. Hence, the success of crystal detectors was substantially enhanced by the advent of radar during World War Two (and, as a result, that of semiconductors). Because to this, a lot of work was done during the war to enhance the semiconductors—primarily silicon and germanium—used in crystal detectors. Russell Ohm at Bell Labs discovered around the same time that these materials could be "doped" with minuscule amounts of foreign "impurity" atoms to provide exciting new features.

Although very many specialised diodes are used in the electronics industry to perform specific functions, the standard silicon or germanium PN junction diodes are used in even greater quantities. They are capable of acting in a wide range of capacities, including those of signal diode, switching diode, power diode, and high voltage diode. Many common PN junction diodes built of silicon or, more frequently, germanium may carry out these and a wide range of other duties.

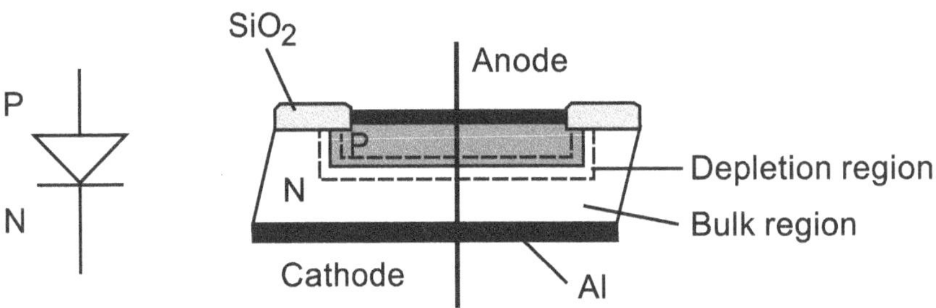

Fig. 7.1 Diffusion junction diode

7.3 DIODE SPECIFICATIONS

When using any diode it is necessary to ensure that it will meet the requirements for the job intended for it. There are a number of important specifications, some of which are more important than others dependent upon the type of role intended for the diode.

7.3.1: Semiconductor material : One of the major elements that will determine the properties of a diode is the material from which it is made. Most general purpose and rectifier diodes are made from silicon.

This provides the finest overall performance at a reasonable cost. Several applications might make use of silicon. More specialised diodes typically make use of different materials.

7.3.2 Forward voltage drop

While any diode will have a certain voltage drop because it is a semiconductor diode, and generally this is taken as about 0.6 volts for silicon and 0.2 volts for germanium, this is dependent upon the current flowing. There is also a resistive drop in the diode that adds to this. Hence, there will be a general voltage decrease for any current. While this may not be important for some circuits, it may be for others, especially where current levels start to rise. For some applications it may mean that enough heat is dissipated for the diode to need a heartsick.

7.3.3 Leakage current

The diode should not conduct any current when it is reverse biased. Unfortunately this is not the case, because a small amount does flow. (This current flows because there are what are termed minority carriers in the semiconductor). The level of the leakage current increases as the temperature rises. The diode leakage current is specified at a certain reverse voltage and temperature and is normally only measured in micro amps or pico amps. Silicon is much better than germanium.

7.3.4 Junction capacitance

Although varactor or varicap diodes have been optimised for their junction capacitance properties, all diodes exhibit capacitance across the junction. The level of capacitance falls as the reverse bias is increased. Accordingly any junction capacitance is specified for a given level of reverse bias. While the amount of junction capacitance is not important for power rectifier diodes, it is important for those operating at higher frequencies. Low capacitance diodes are hence readily available. The 1N4149, for instance, is the 1N4148's low capacitance equivalent.

7.3.5 Peak Inverse Voltage

This is the highest reverse voltage that a diode can withstand. It is one of the key specifications for high voltage diodes and power diodes used in power supplies.

Diodes used in power electronics applications are generally required to have special characteristics, these are:

- · high breakdown voltage and current carrying capacity
- · Short switching times and quick rise and fall times of the current; ·
- · Reverse recovery is insufficient. That is, there is very little charge removal at turn OFF.
- · Low voltage drop when conducting.

Unfortunately, it is not possible to achieve this entire criterion with one single style of diode and thus a number of different types of power diode are available for various applications. It is up to the circuit designer to judge which component is best suited for a particular application this will often result in a conflict between what is required and what is available and it is here the circuit design can be very important.

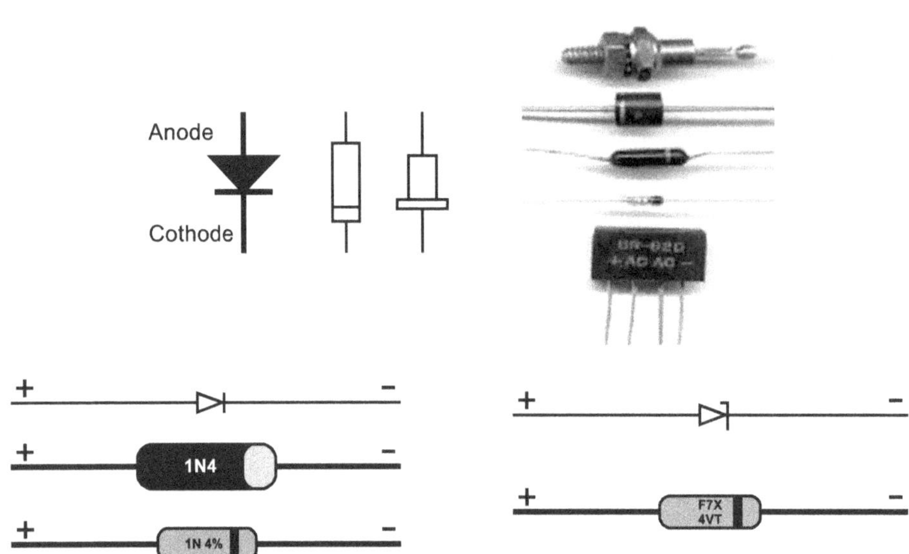

Fig. 7.2 circuit symbol polarity of different diodes

7.4   JUNCTION DIODE BEHAVIOR

The capacity of a junction diode to conduct an electric current in only one direction is its most significant characteristic. If the diode is connected to a simple circuit consisting of a battery and a resistor, the battery can be connected in either of two ways as shown in Figures 7.3 a,b.

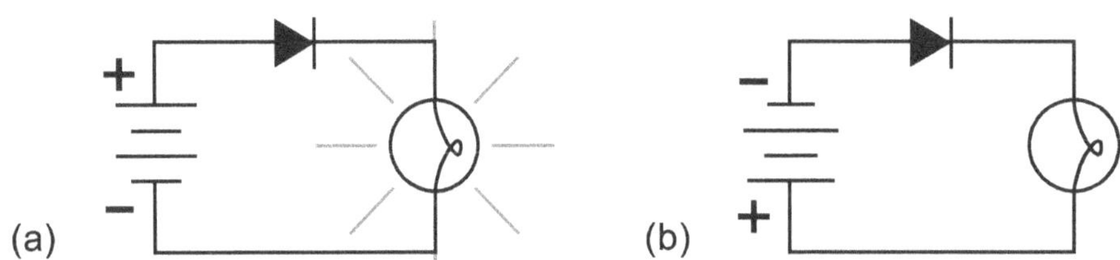

Fig. 7.3 Current flow in PN Junction diode

When the positive terminal of the battery is connected to the p-type region of the p-n junction, current will flow. It is claimed that the diode is biased forward. On the other hand, the p-n junction almost entirely obstructs the current flow when the battery terminals are switched around. The phrase for this is reverse bias. When a diode is totally unplugged, it is said to be open-circuited since no current can flow through it.

A graph showing the relationship between the applied voltage and the measured current is a useful tool for demonstrating this idea.

This voltage-current relationship (characteristic curve) for a typical p-n junction diode is depicted in Figure 7.4.

67

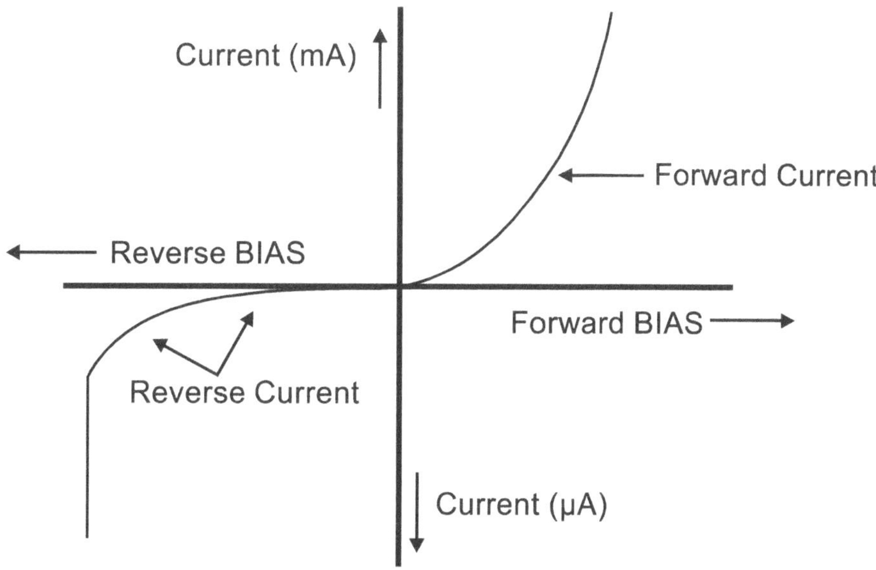

Fig. 7.4 VI-Characteristic for PN-junction diode

Points to think:

1. One piece of advice is that a diode is an electrical component that controls how current flows.

2. A diode is said to be forward-biased if voltage is applied across it in such a way that it allows current to flow.

3. When a voltage is applied across a diode in a way that prohibits current flow, the diode is said to be reverse-biased.

4. The voltage that is decreased across a conducting, forward-biased diode is known as the forward voltage. The forward voltage of a diode is fixed by the chemical composition of the P-N junction and only minimally varies with variations in forward current and temperature.

5. A forward voltage of roughly 0.7 volts is present in silicon diodes.

6. About 0.3 volts is the forward voltage of germanium diodes.

7. Peak Inverse Voltage, also referred to as PIV rating, is the maximum reverse-bias voltage that a diode can withstand before "breaking down."

7.5   RATING OF DIODES

The Peak Inverse Voltage, or PIV, of a diode is its maximum reverse-bias voltage rating and can be obtained from the manufacturer. Similar to forward voltage, a diode's PIV rating changes with temperature. However, unlike forward voltage, PIV rises with higher temperatures and falls as the diode cools.

A generic "rectifier" diode's PIV rating is typically at least 50 volts at ambient temperature. For affordable pricing, diodes with PIV ratings in the tens of thousands of volts are readily accessible.

Ratings of Diode

For a wide range of semiconductor components, it is data sheets. In the product manual, the manufacturer provides specifications for these components.

Figures for the following parameters are usually included in a diode datasheet:

1. Maximum repetitive reverse voltage, also known as VRRM, is the highest voltage that a diode can tolerate when operating in reverse-bias mode and is measured in pulses. This amount should ideally never end.

2. The greatest voltage the diode can withstand when operating in reverse-bias mode is known as the maximum DC reverse voltage, abbreviated as VR or VDC. This amount should ideally never end.

3. Maximum forward voltage = VF, often stated at the rated forward current of the diode. A zero value for this number is ideal.

4. The highest average forward current the diode is capable of conducting in the forward bias mode is given by the formula $I_F V_F/R$), where V_F is the maximum average voltage. This is essentially a thermal limit: how much heat can the PN junction withstand? The fundamental question is how much power the PN junction can dissipate because forward voltage depends on both current and junction temperature, and dissipation power is equal to current (I) times voltage (V or E). Ideally, this amount should never be exhausted.

The maximum peak current that a diode can carry in the forward bias mode is called the greatest (peak or surge) forward current. Once more, the diode junction's thermal capacity acts as a restriction for this rating, which is ordinarily much higher than the average current rating due to thermal inertia (the fact that it takes a finite amount of time for the diode to reach maximum temperature for a given current). This number should ideally be limitless.

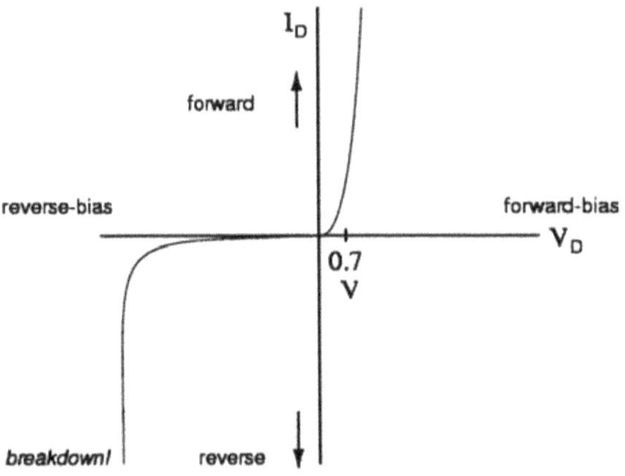

Fig. 7.5 Si diode showing knee at 0.7 V forward bias and reverse breakdown

7.6 DIFFERENT TYPES OF DIODE

The most popular and affordable diodes have a forward voltage drop of roughly 0.65 volts and are made of silicon. A forward voltage drop of roughly 0.1 volt is present in germanium diodes. However the cost of germanium diodes is often substantially higher than that of silicon diodes.

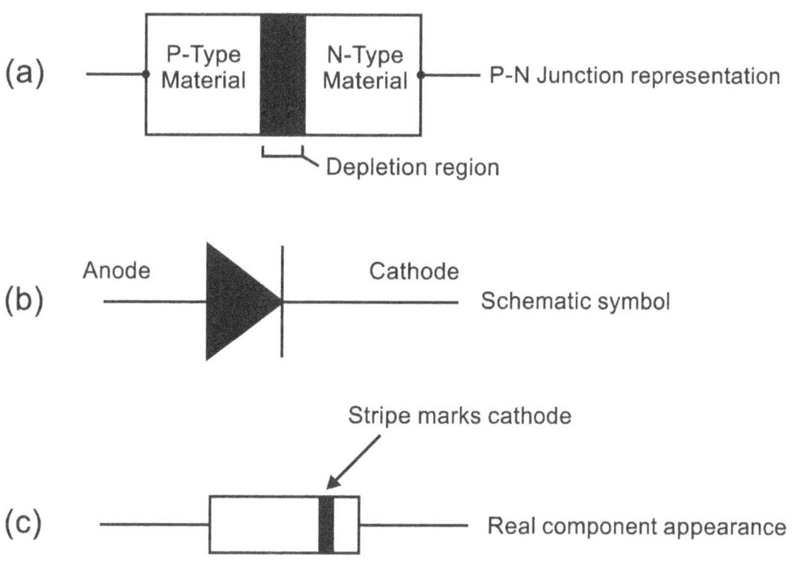

Fig. 7.6 : SI-GE diode a) P-N Junction b)Schematic symbol c) physical appearance

7.6.1 Zener Diodes

To keep a fixed voltage, zener diodes are employed. They are made to "breakdown" in a trustworthy and non-destructive manner so that they can be used backward to keep a constant voltage across their terminals.

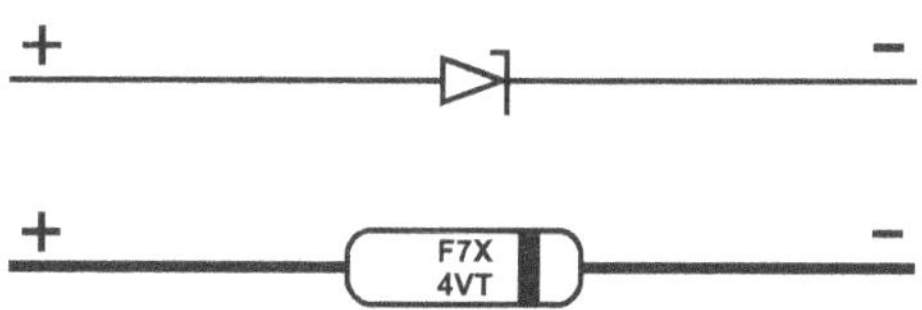

Fig. 7.7 : Zener diode Circuit, Schematic symbol

7.6.2 LEDs (Light Emitting Diode)

All diodes produce some light when biassed forward. A unique semiconductor, such as gallium arsenide phosphide, is used in the construction of LEDs to maximise this light output. Until their current limit is exceeded, LEDs don't often burn out like light bulbs do.\

70

Fig. 7.8 : LEDs Circuit, Schematic symbol

The voltage on the positive leg of an LED is around 1.4 volts greater than the voltage on the negative side when current is flowing through it (this varies with LED type infrared LEDs have a lower forward voltage requirement, others may need up to 1.8 V There is relatively little resistance to control the current, thus a resistor must be connected in series with the LED to prevent damage.

Moreover, keep in mind that LEDs can function as photo diodes (although this only works in extremely bright situations due to their low sensitivity).

SPECIAL-PURPOSE DIODES : Varactor diodes , Photo diodes , Schottky diodes , PIN diodes , Tunnel diodes, Laser diodes, Solar cells etc.

7.7 TESTING OF DIODE

Determining polarity is a crucial ability for an electronics expert to possess. As a diode is really nothing more than a one-way electrical valve, we should be able to confirm this using an ohmmeter, as shown in Figure 7.9. The metre should display a very low resistance at when connected one way across the diode (a). It should display an extremely high resistance at (b) ("OL" on some digital metre models) when connected the other way across the diode.

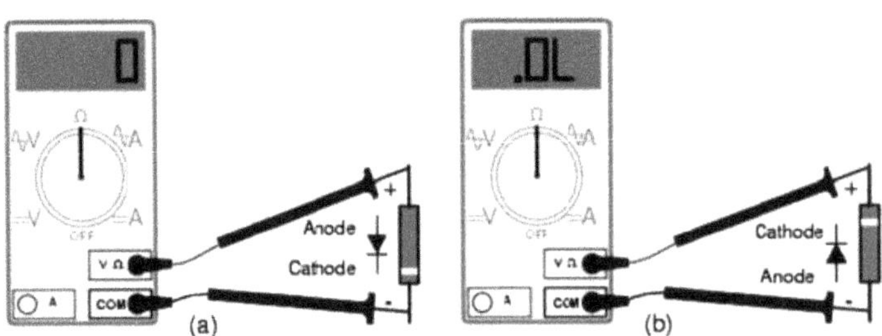

Fig. 7.9 : Diode Testing for polarity using Ohmmeter

Black lead is the cathode and red lead the anode, and low resistance indicates a forward bias. Reverse bias is indicated by increased resistance in reverse leads.

Some manufacturers of digital multi-meters give their instruments a particular "diode check" feature that shows the diode's actual forward voltage drop in volts rather than its "resistance" value in ohms. As depicted in fig. 7, these metres function by applying a tiny current to the diode and detecting the voltage drop between the two test leads. Instead of showing a low resistance, a 10 Meter with the "Diode check" feature shows the forward voltage loss of 0.548 volts.

For silicon and germanium, the "normal" drop will be 0.7 volts and 0.3 volts, respectively, however the forward voltage reading obtained with such a metre will typically be lower.

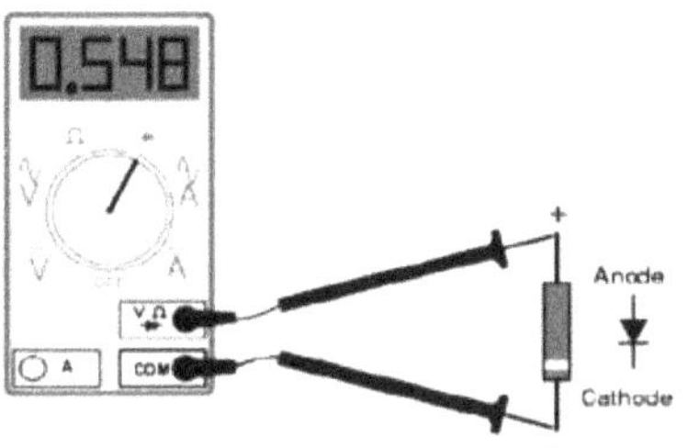

Fig. 7.10 : Diode Testing for polarity using Voltmeter

7.8 RECTIFIER

7.8.1 Half Wave Rectifier

Diodes are most frequently used for rectifying. Rectification is the conversion of alternating current (AC) to direct current (DC) in its most basic form. A piece of machinery that only allows one-way electron flow is used in this situation. As we have seen, a semiconductor diode accomplishes this. The half-wave rectifier is the simplest basic type of rectifier circuit. It only allows one half of an AC waveform to pass through to the load (a), as seen in fig. 7.11. It is used for both bright lights and dimmable lamps.

7.8.2 Full Wave Rectifier

Even though we can use both sine wave cycles (AC), a different rectifier circuit arrangement must be used. A circuit like this is referred known as a full-wave rectifier. One kind of full-wave rectifier, as seen in Fig. 7.11, makes use of two diodes and a transformer with a center-tapped secondary winding (b).

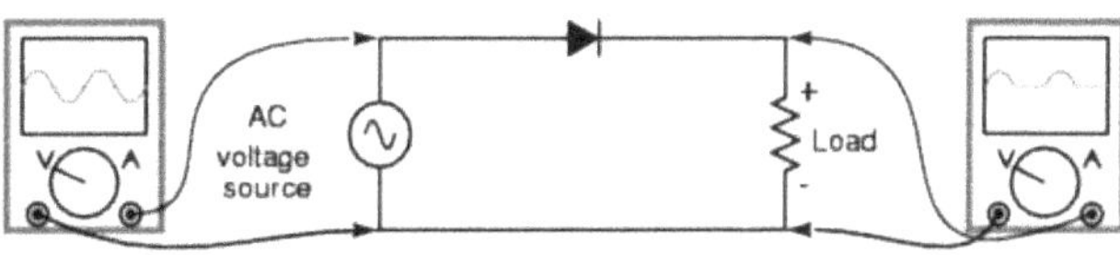

Fig. 7.11 (a) : Half wave rectifier

When the polarity is positive (+) on top and negative (-) on the bottom, visualize the AC voltage throughout the first half of the cycle. At this time, the load "sees" the top and bottom halves of the first half of the sine wave. Only the top diode is conducting while the bottom diode is obstructing current. . As seen in Fig. 7.11, only the upper half of the transformer's secondary winding carries electricity during this half-cycle (b).

The next half-cycle is when the polarity of the AC is changed. The other diode and the other half of the secondary winding of the transformer are now carrying current, while the components of the circuit that were previously carrying current during the previous half-cycle are now inert. The load continues to "see" a sine wave with positive on top and negative on bottom with the same polarity. These centre-tap rectifiers are useful in applications where less power is required.

7.8.3 Bridge Wave Rectifier

It is constructed using a four-diode bridge arrangement. This design is known as a full-wave bridge for obvious reasons.

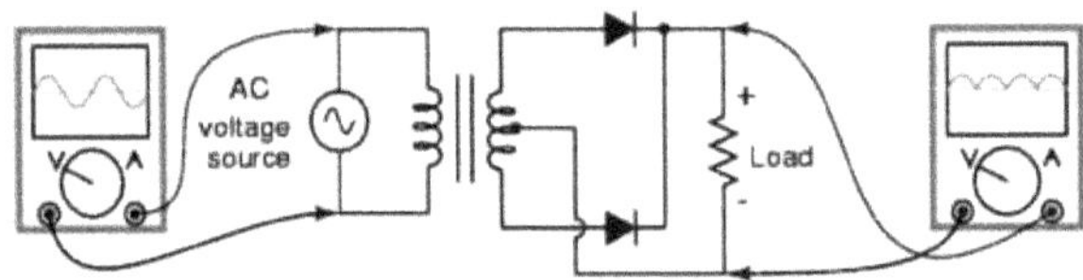

Fig. 7.11 (b) : Full wave rectifier

For the positive half-cycle of the AC, the current directions for the full-wave bridge rectifier circuit are as indicated in Fig. 7.12.

Observe that the current always flows through the load in the same direction, irrespective of the polarity of the input. In other words, at the load, the negative half-cycle of the source is a positive half-cycle. Two diodes are connected in series to allow current to flow in both polarities.

Due to two diode losses, the source voltage is subsequently lost in the diodes (0.72=1.4 V for Si). This is less advantageous than a full-wave center-tap configuration.

Only very low voltage power sources experience this problem. The advantage of bridge rectifier circuits is that they may be easily expanded into polyphase systems (three-phase or six-phase AC source).

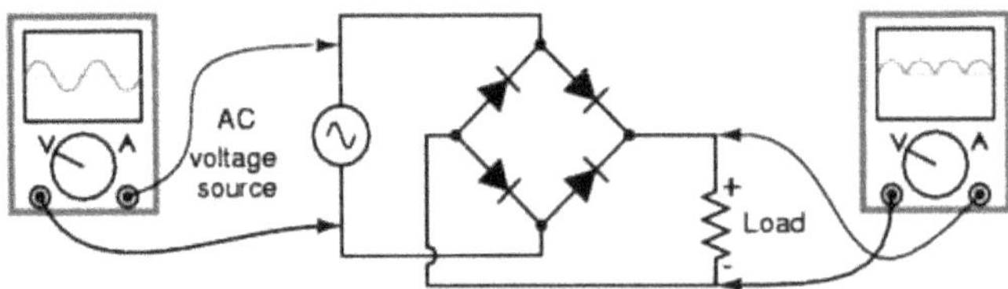

Fig. 7.12 Full wave bridge rectifier

CHAPTER-8

Transistor

8.1 BELL TELEPHONE LABORATORIES

Bell Laboratories was where the transistor was created in 1948. The majority of the earliest entertainment technology utilised germanium-type transistors. The silicon transistor's popularity skyrocketed after its invention. The transistor's early benefits were its comparatively low power requirements at low voltage levels, which made it possible to produce portable entertainment devices on a large scale. It's interesting to note that the transistor industry has grown with the battery sector.

8.2 LAYERS OF A TRANSISTOR

Three layers of silicon or germanium semiconductor material make up a transistor. Each layer is given impurities to produce a certain electrically positive or negatively charged behaviors. The letters "P" and "N" stand for positive and negative charges, respectively. In terms of the layer configuration, transistors can be either NPN or PNP. The leads have the letters B, C, and E on them (E). Three leads on transistors must be linked in the proper order. Please exercise caution because switching on could rapidly harm a transistor that is improperly connected. Refer to figures 8.1 and 8.2 for your information.

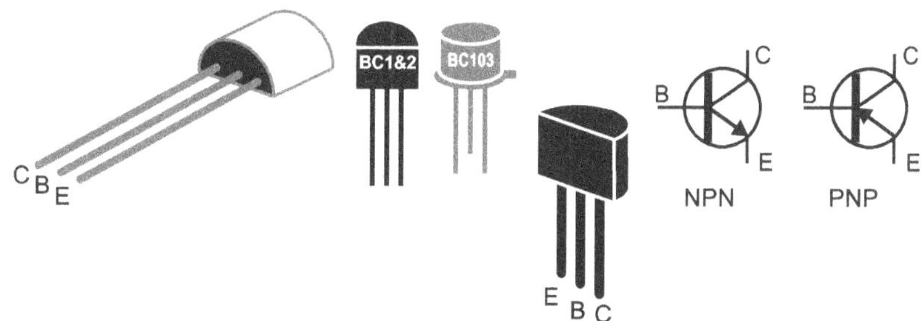

Fig. 8.1 Transistors Schematic and circuit symbol

8.3 TRANSISTOR - A BASE OF ELECTRONICS

All of today's electronics depend on transistors as fundamental parts. To turn everything on and off, we can merely utilise these straightforward switches. Despite being straightforward, they are the most crucial electrical component. For instance, a Pentium CPU is almost entirely made of transistors. The number of transistors on a Pentium chip is around 3.5 million. Despite being smaller than the ones we'll be using, the ones in the Pentium perform the identical function.

8.4 FAMILY OF A TRANSISTOR

In general, transistors belong to the bipolar transistor category, and they can either be NPN bipolar transistors, which are more prevalent, or PNP transistor types, which are less prevalent. Another kind is the FET transistor, which has an input impedance that is naturally high and behaves somewhat like valves. Contemporary field effect transistors, sometimes known as FETs, have several incredibly durable transistor devices, like JFETS and MOSFETS. Now, we must merely mention the transistor.

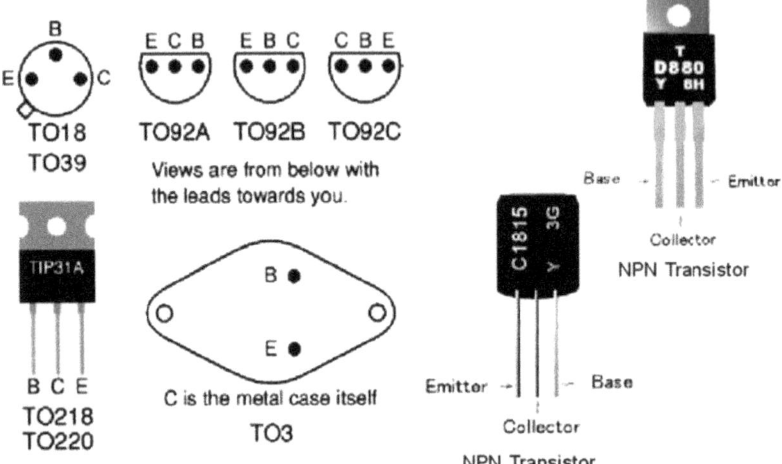

Fig. 8.2 Transistor leads for some common case styles

8.5 APPLICATION OF A TRANSISTOR

An amplifier or electrically controlled switch are two popular uses for transistors, which are semiconductor devices. The transistor is the fundamental component of all modern electronic devices, including computers, smart phones etc.

The transistor performs a wide range of analogue and digital tasks, including amplification, switching, voltage control, signal modulation, and oscillators, because to its quick response and accuracy. Individually packaged transistors or those that are a component of an integrated circuit can have over a billion of them packed into a very small space.

Transistors are used in analogue circuit components such as amplifiers (direct current amplifiers, audio amplifiers, and radio frequency amplifiers). In addition to being employed in analogue circuits, where they serve as electronic switches, transistors are also used in digital circuits. However, they are practically never used as discrete devices in these circuits; rather, they are almost always a part of monolithic integrated circuits. RAM, digital signal processors, logic gates, and microprocessors are all examples of digital circuits (DSPs).

8.6 TRANSISTOR TESTING

Heat from soldering or incorrect use in a circuit can harm transistors. There are two simple tests you can perform if you think a transistor could be damaged:

8.6.1 Transistor Testing by Multimeter

Use a simple tester or a multimeter.

To examine each pair of leads for conduction, use a (battery, resistor, and LED). Configure the diode test function on a digital multimeter and the low resistance range on an analogue multimeter.

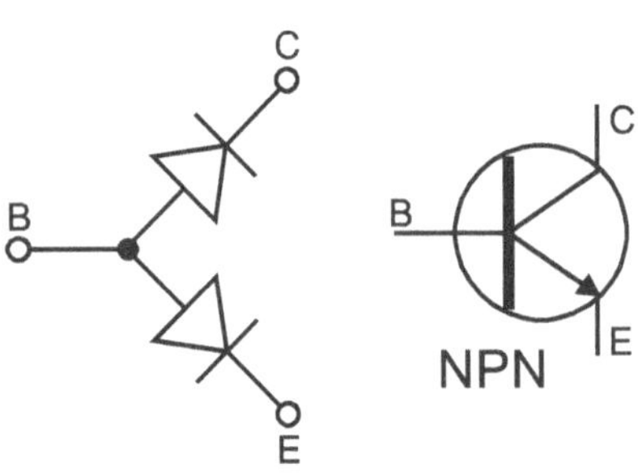

Fig. 8.3 Testing of NPN Transistor

Test each pair of leads in both directions (a total of six tests):

The base-emitter junction (BE junction) need to operate like a diode and conduct in just one direction.

The base-collector junction (BC junction) need to operate like a diode and conduct in only one direction.

Neither way should be conducted by the collector-emitter (CE).

Figure 8.3 illustrates the behaviour of the junctions in an NPN transistor. In a PNP transistor, the diodes are switched around, but the same test method can be applied.

8.6.2 A simple switching circuit Testing

Attach the transistor to the circuit, which employs the transistor as a switch, as shown on the right. It is not necessary to have a specific supply voltage; anything between 5 and 12V will do. This circuit can be quickly constructed, for instance, using a breadboard. While testing the transistor, be sure to include the 10k ohm resistor in the base connection to avoid damaging it. If the transistor is functioning properly, the LED should turn on when the switch is depressed but not when it is released. Use the same setup to test a PNP transistor, but switch the LED and supply voltage around.

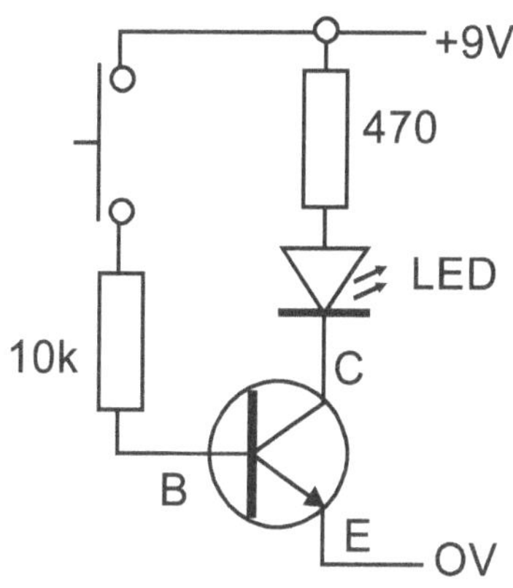

Fig. 8.4 Testing of NPN Transistor with circuit

8.7 WORKING OF TRANSISTOR

8.7.1 The NPN Transistor

We would typically expect that the base voltage will be between 0.65 and 0.7 volts positive with respect to the emitter, that electrons will move from the emitter to the base, and that they will then leave the device at that point because we are already familiar with how a single-junction device, the diode, operates. We would anticipate no current to pass through the collector junction when it is reverse biased.

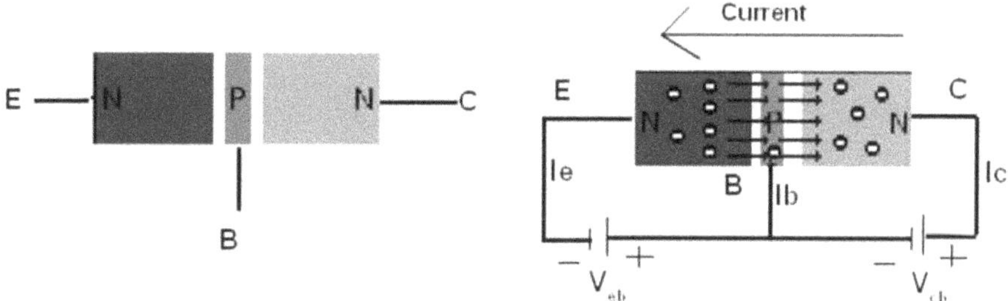

Fig. 8.5 Biasing of NPN Transistor

It is true that the forward bias on this junction draws electrons from the emitter into the base, but once inside, their forward velocity carries them over the majority of the base region and into the depletion zone around the collector junction. These electrons are then drawn into the collector region by the larger positive collector voltage, which is present there. Note that the reverse-biased junction might be crossed by a leakage current because electrons in the P-type base region are minority current carriers.

Although most of the current is redirected through the collector, a tiny amount of current does still exit the device through the base contact. The modest base bias current thus regulates the enormous collector current. The collector current will change significantly more if a tiny changing current is applied to the base along with the bias. As a result, this gadget has the ability to both control and magnify diverse signals. Whereas in Figure 8.5

Emitter=E   Collector=C   Base=B   Collector current=Ic

Emitter current=Ie   Base current=I_b h_{fe} = Current gain(DC)

I)   To control the base current I_b and guard against transistor damage, it is frequently necessary to connect a resistor in series with the base connection.

ii) The maximum collector current Ic rating for transistors.

iii) As depicted in fig. 8.5, the direction of the current will be the polar opposite of the flow of electrons

iv) The collector-emitter voltage V_{EC} is almost zero volts when a transistor is saturated.

v)   The supply voltage and the external resistance in the collector circuit, not the transistor's current gain, control the collector current Ic when a transistor is saturated. As a result, the current gain h_{fe} for a saturated transistor is lower than the ratio I_c/I_b.

vi) The emitter current I_e is equal to I_c plus I_b, but since I_c is significantly bigger than I_b, I_e is essentially equal to I_c.

8.7.2 PNP Transistor

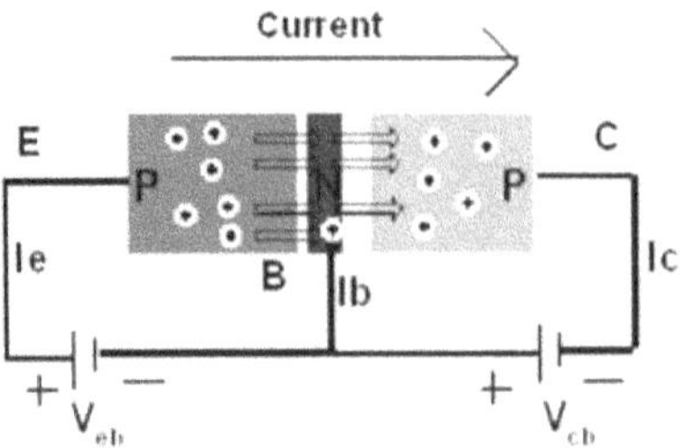

Fig. 8.6 Biasing of PNP Transistor

The functionality and general characteristics of this device are otherwise identical to those of the one shown in fig. 8.6.

We refer to the two types of transistors according to the sequence in which the various regions emerge.

i)   The collector is always reversed biased while the emitter is always forward biased.

ii) In the emitter part, holes will travel in a direction towards the junction.

iii) The hole will be discharged by an electron moving from the Vcb negative terminal to the emitter, or collector.

iv) The current will flow from the emitter to the collector.

8.8 TRANSISTOR AS AN AMPLIFIER

A huge change in collector current is produced for a little change in base current by the basic transistor amplifier, which amplifies. Due to the load resistor's response to these significant variations in collector current, which in turn create substantial variations in the output voltage, this action leads in voltage amplification.

Amplifiers typically use transistors. Some transistor circuits are current amplifiers with low load resistance; other transistor circuits are voltage amplifiers with high load resistance; still other transistor circuits are power amplifiers.

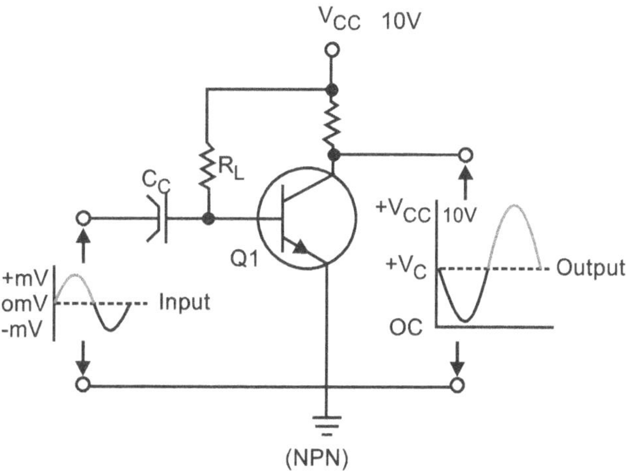

Fig. 8.7 NPN Transistor as an Amplifier

The most typical configuration for transistor amplifiers is the common emitter configuration, which is well-suited to voltage amplification.

Amplification is the process of boosting a signal's power.

The device that amplifies sound without significantly changing the original signal is called an amplifier.

Base-current bias (fixed bias), self-bias, and combination bias are the three types of BIAS that are utilized to correctly bias a transistor.

The most popular type of bias is combination bias because it reduces some of the drawbacks of base-current bias and self-bias while also enhancing circuit stability.

Figure 8.7 demonstrates the removal of the emitter-base battery and the placement of the bias resistor RB between the collector and the base. The necessary forward bias for the emitter-base junction is provided by resistor R_B. In the emitter-base bias circuit, current travels from the ground to the emitter, out the base lead, through R_B to R_L, and back again.

When Q1 is suitably biased, the circuit operates constantly, whether or not there is an input signal. As the direct current passes through Q1 and V_C, it creates the collector voltage (Vcc), in addition to base bias, in the circuit. Take note of the output graph's collector voltage. The output signal starts at the V_C level since it exists in the circuit even in the absence of an input signal and then either rises or falls. Quiescent voltages and currents are these dc values that exist in the circuit without the application of a signal (the quiescent state of the circuit).

To prevent the collector from being fully affected by the collector supply voltage, the collector load resistor R_L is added to the circuit. This enables the collector voltage (V_C) to fluctuate in response to an input signal, enabling the transistor to amplify voltage. The voltage on the collector would always be equal to V_{CC} in the absence of.

Only a few tenths of a volt of positive bias will be felt on the transistor's base due to the base circuit's extremely low current (a few hundred microamperes) and low forward resistance. The large positive voltage on the collector, ground on the emitter, and this sufficient voltage on the base, however, allow the transistor to be biased as intended.

The input signal resists the forward bias during the negative alternation of the input. Because to the reduction in base current, the currents flowing through the emitter and collector are also reduced. The voltage across the transistor increases along with the output voltage when the current through R_L is reduced, resulting in a reduction in the voltage drop across the transistor.

As a result, the output of a negative input alternation is a positive input alternation of voltage that is bigger than the input but has the same sine wave properties.

In conclusion, the slight change in base current resulted in a huge change in collector current, which amplified the input signals in the circuits before it. The voltage was amplified by connecting resistor R_L in series with the collector.

8.9 CONFIGURATION OF TRANSISTOR

It refers to the specific manner a transistor is wired up in a circuit. There are three alternative ways to connect a transistor: common emitter (CE), common base (CB), and common collector (CC).

8.9.1   Setup of a common emitter (CE)

Since it offers good voltage, current, and power gain, it is the configuration that is employed the most commonly in real-world amplifier circuits, as seen in fig. 7.8. The emitter is the element that is "common" to both the input and output of the CE since the input is applied to the base-emitter circuit and the output is drawn from the collector-emitter circuit. The lone configuration, the CE, which offers a phase reversal between input and output signals, distinguishes itself from the other variants.

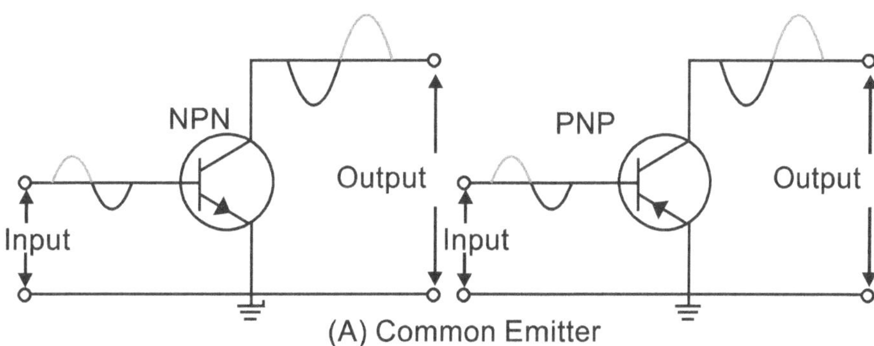

Fig. 8.8 Transistor CE Amplifier

8.9.2   Setup of a Common-Base Configuration (CB)

Since it has a high output resistance and a low input resistance, its primary application is impedance matching. Moreover, it now has a gain of under 1.

In fig. 8.9, the emitter receives the input, the collector receives the output, and the base is the element that is present in both the input and the output.

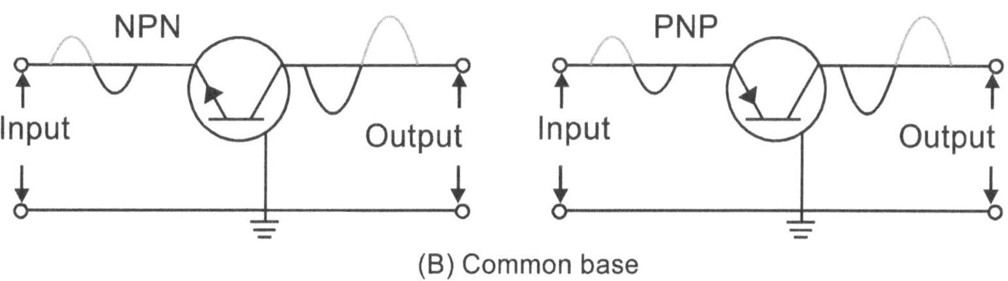

Fig. 8.9 Transistor CB Amplifier

8.9.3   Setup of a Configuration of a common-collector (CC)

It is particularly helpful in switching circuits and is used as a current driver for impedance matching. The CC, also known as an emitter-follower, is the cathode follower of an electron tube. Each has a low output impedance and a high input impedance.

The input and output in Fig. 8.10 are applied to the base and emitter, respectively, with the common element between the two being the collector.

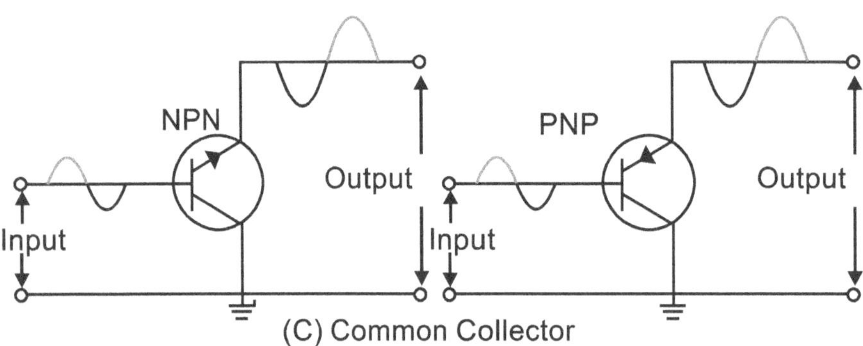

Fig. 8.10 Transistor CC Amplifier

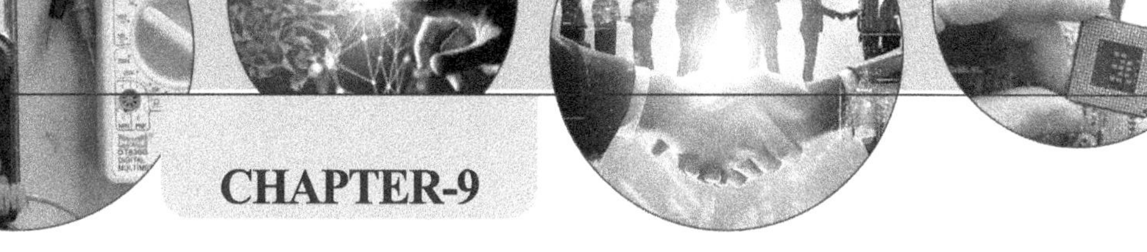

CHAPTER-9

Specialized Devices

9.1 ZENER DIODE

The voltage drop across a forward-biased diode will be pretty constant throughout a wide variety of power supply voltages if a resistor and diode are connected in series with a DC voltage source as shown in Figure 9.1. (a).

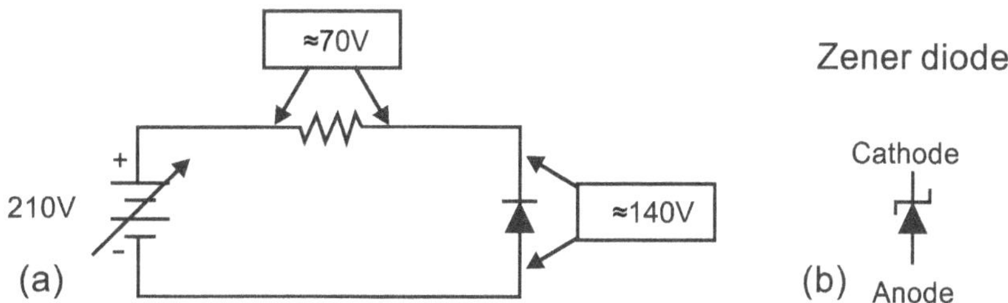

Figure 9.1 (a) Reverse biased Around 140V causes the Si small-signal diode to fail.

In the circuit above, the voltage of the power supply, the series resistor, and the diode's voltage drop—which, as we know, doesn't deviate significantly from 0.7 volts—all serve to limit the diode current. The voltage drop of the resistor and the diode would both increase almost proportionally if the power source voltage were raised. In contrast, a decline in power supply voltage would cause the voltage drop across a resistor to drop by almost the same amount, and the voltage drop across a diode to drop by just a tiny amount.

Voltage control is a beneficial diode feature to take use of. We may set up a circuit in the manner depicted and connect the circuit that needs a constant voltage across the diode, providing it with a constant 0.7 volts.

This would undoubtedly perform, however the majority of real-world circuits of any kind need a power supply voltage greater than 0.7 volts to operate correctly. Connecting more diodes in series would allow their separate forward voltage dips of 0.7 volts apiece to build up to a bigger sum, increasing our voltage regulation point. For instance, the regulated voltage would be 10 times 0.7, or 7 volts, if we had ten diodes connected in series.

Unluckily the majority of the time when typical rectifying diodes "break down," they do so destructively. Yet, it is feasible to create a unique kind of diode that can withstand breakdown without fully failing. Figure 9.1 depicts the sign for this kind of diode, known as a zener diode (b).

9.2 VARICAP OR VARACTOR DIODE

An insulating depletion area arises between the two semi conductive layers of a diode, which is a reverse biased feature. By adjusting the reverse bias, the breadth of the depletion region can be altered in many diodes. Capacitance varies as a result. A varactor or a varicap diode is a device that has changeable capacitance. Figure 9.1 below depicts the schematic symbols, one of which is packaged as a common cathode dual diode.

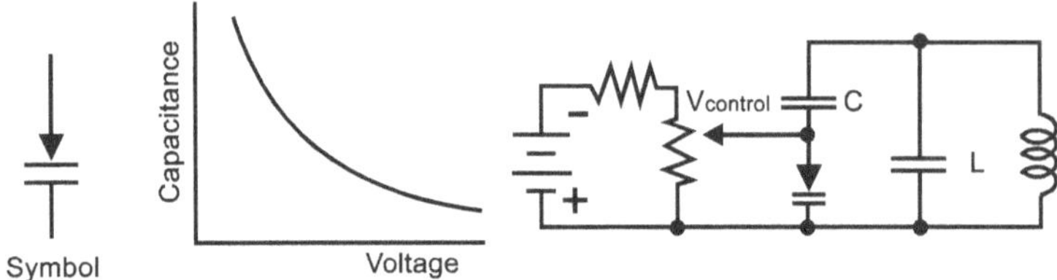

Fig. 9.2 Vractor diode symbol, graph, resonant circuit.

A control voltage, Vcontrol, may be used to change the frequency of a varicap diode if it is a component of the resonant circuit shown in Fig. 9.2 above. Inductor L protects Vcontrol from being shorted out by adding a large capacitance with a low Xc value in series with the varicap. The frequency of the resonant circuit is not significantly affected by the series capacitor as long as it is large. To determine the centre resonant frequency, utilise C optional. The frequency can then be changed using Vcontrol at this point.

9.3 THE JUNCTION FIELD EFFECT TRANSISTOR (JFET)

A semiconductor bar with ohmic connections at one end and strongly doped regions on its opposing sides may be the basic construction of a junction field effect transistor (JFET). An n-channel JFET is one in which the semiconductor bar is constructed from n-type material. If the material used to make the bar is p-type, the JFET is p-channel.

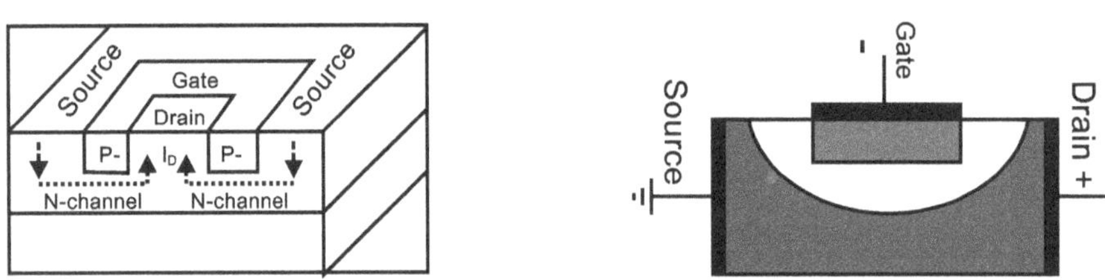

Fig. 9.3 : JFET Geometry structure, schematic symbol

The JFET's source and drain are represented by the terminals at either end of the bar. The severely doped regions connecting to the bar's sides act as the JFET's gate. Naturally, the gate regions are doped in a way that is opposite to the channels doping, resulting in a p-n junction between the channel and the gate regions.

D = Drain

S = Source

G = Gate

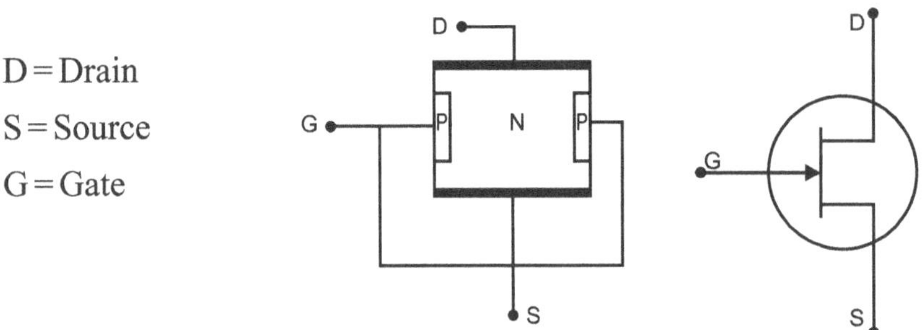

Fig. 9.4 : JFET schematic, circuit symbol

Current made up of of majority carriers—holes for a p-channel and electrons for an n-channel-is made to flow through a JFET's channel by putting a voltage across its source and drain.

Applying a gate voltage Vgs that reverse biases the p-n junction created by the gate with respect to the source regulates the amount of current flowing through the channel. The p-n junction is increasingly reverse-biased and the depletion zone across the channel is broader as Vgs increases.

The smaller channel that arises from the larger depletion zone restricts the passage of current across the channel. For any given voltage across the source and drain, changing Vgs alters the current through the channel.

Because only the majority carriers pass through the channel of a MOSFET or JFET, FETs are sometimes known as unipolar transistors.

9.4 UNIJUNCTION TRANSISTOR(UJT)

The unijunction transistor is one of the strangest semiconductor devices now in use. Despite having three terminals, it only has one PN junction. Despite the fact that it cannot magnify an applied signal, it can be utilised as the active component of an oscillator circuit.

A typical UJT's physical design is depicted in fig.9.5. A bar of N-type silicon with electrical connections at either end and an aluminium wire attached to a spot along the silicon bar's length make up the device. Aluminium at the bonding point induces a P-type area in the silicon bar, resulting in the formation of a PN junction.

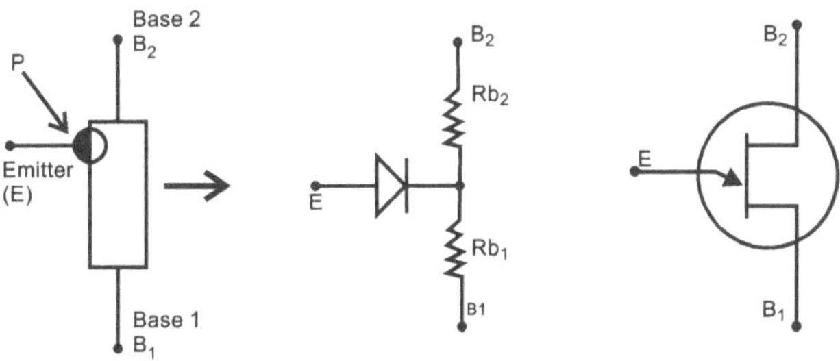

Fig. 9.5 : UJT schematic, circuit symbol

When in use, the two bases are subjected to an appropriate bias voltage, with B_2 being rendered positive in relation to B_1. The applied voltage will be dispersed uniformly down the length of the N-type bar because it is resistive, allowing just a little amount of current to pass through it. The junction will be reverse biased and there won't be any emitter current if the emitter is grounded at the beginning. There is no change as the emitter voltage rises until the junction abruptly turns forward biased.

 Until its voltage hits Vp, the emitter terminal does not inject current into the base area. When Vp is attained, the base circuit conducts, causing positive and negative pulses to emerge at terminals B_1 and B_2, respectively. The UJT is the ideal pulse trigger because it has a negative resistance area, a low emitter current, and a high output pulse current at terminals B_1 and B_2.

9.5 SILICON CONTROLLED RECTIFIER (SCR)

One member of the semiconductor family, which also consists of transistors and diodes, is the SCR. Figure 9.6 depicts an SCR's schematic representation and illustration (C,D) Although not all SCRs have the casing shown, most high-power devices do.

Four layers of semiconductor material, ordered PNPN, make up the SCR. View A of fig. 9.6 depicts the structure. Although the SCR functions very similarly to a diode, its principle of functioning is best explained in terms of transistors.

Take the SCR into account as a pair of transistors, one PNP and the other NPN, coupled as depicted in fig. 9.6. (A, B). The gate terminal, G, connects to the P-layer of the NPN triode, while the anode, A, is connected to the upper P-layer and the cathode, C, is a component of the bottom N-layer.

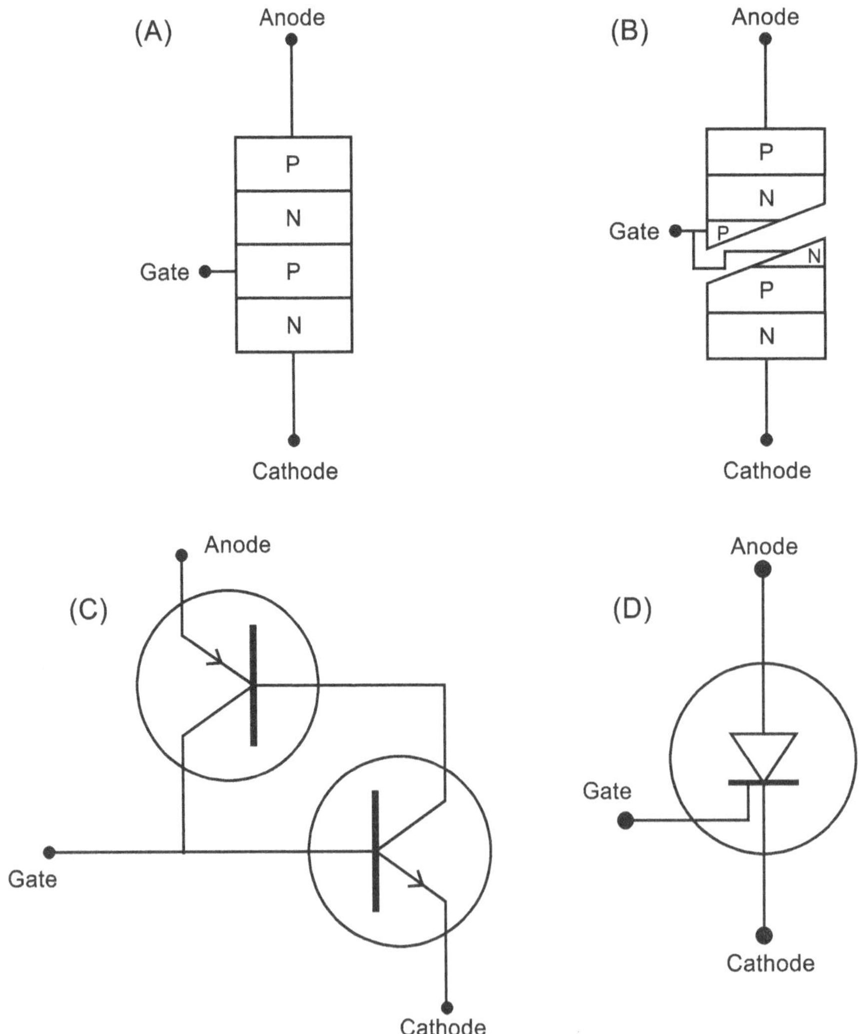

Fig. 9.6 : SCR schematic, equivalent circuit(C) symbol(D)

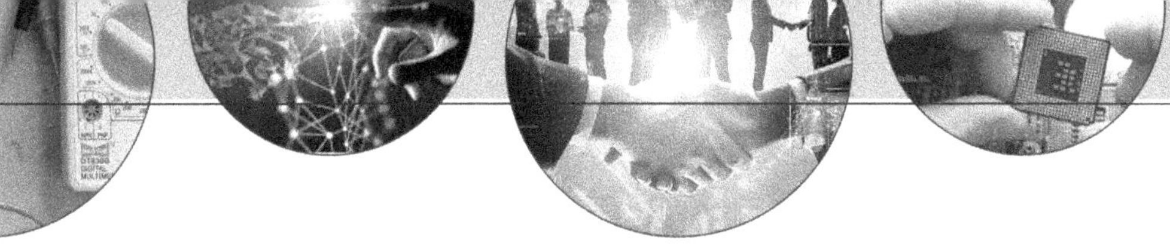

The SCR is a switch that is very quick. A mechanical switch can only be cycled a few hundred times each minute, but certain SCRs can flip 25,000 times per second. These units can be turned on or off in just a few microseconds (millionths of a second). The quantity of power flowing through a switch can be adjusted by varying the time that it is on compared to the time that it is off. The SCR can be utilised easily in control applications since the majority of devices can run on power pulses (alternating current is a special kind of alternating positive and negative pulse). The SCR is used in a variety of devices, including inverters, latching relays, circuit overload protectors, controlled rectifiers, remote switching units, and computer logic circuits.

The SCR automatically shuts off when an ac power supply is utilised since current and voltage zero out every half cycle. Full-wave rectification can be achieved, giving control over the complete sine wave, by utilising one SCR on positive alternations and one on negative. The SCR functions in this application as a controlled ac voltage rectifier, precisely as its name suggests.

The SCR can be used as a rectifier in a variety of situations. In reality, the name of this semiconductor device comes from its numerous uses as a rectifier. A rectifier only allows the positive or negative half of a sine wave to pass through it when alternating current is supplied. The output contains the entirety of each positive or negative half cycle.

Nevertheless, when an SCR is employed, the controlled rectifier can be activated at any point throughout the half cycle, allowing the user to regulate the available dc power from zero to maximum. If continuous direct current is required, adequate filtering can be applied because the output is essentially dc pulses.

Consequently, controlled power can be applied to any dc-operated equipment. Keep in mind that each cycle requires the SCR to be activated at the desired moment.

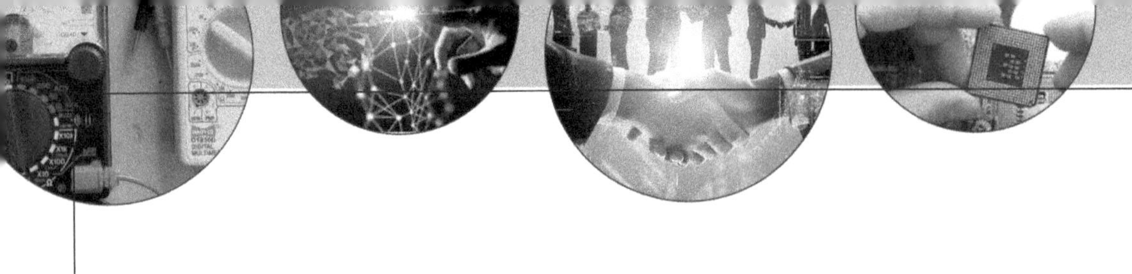

Doing what we do best and later
"something special in the carrier"....

PART-II

DIGITAL ELECTRONICS

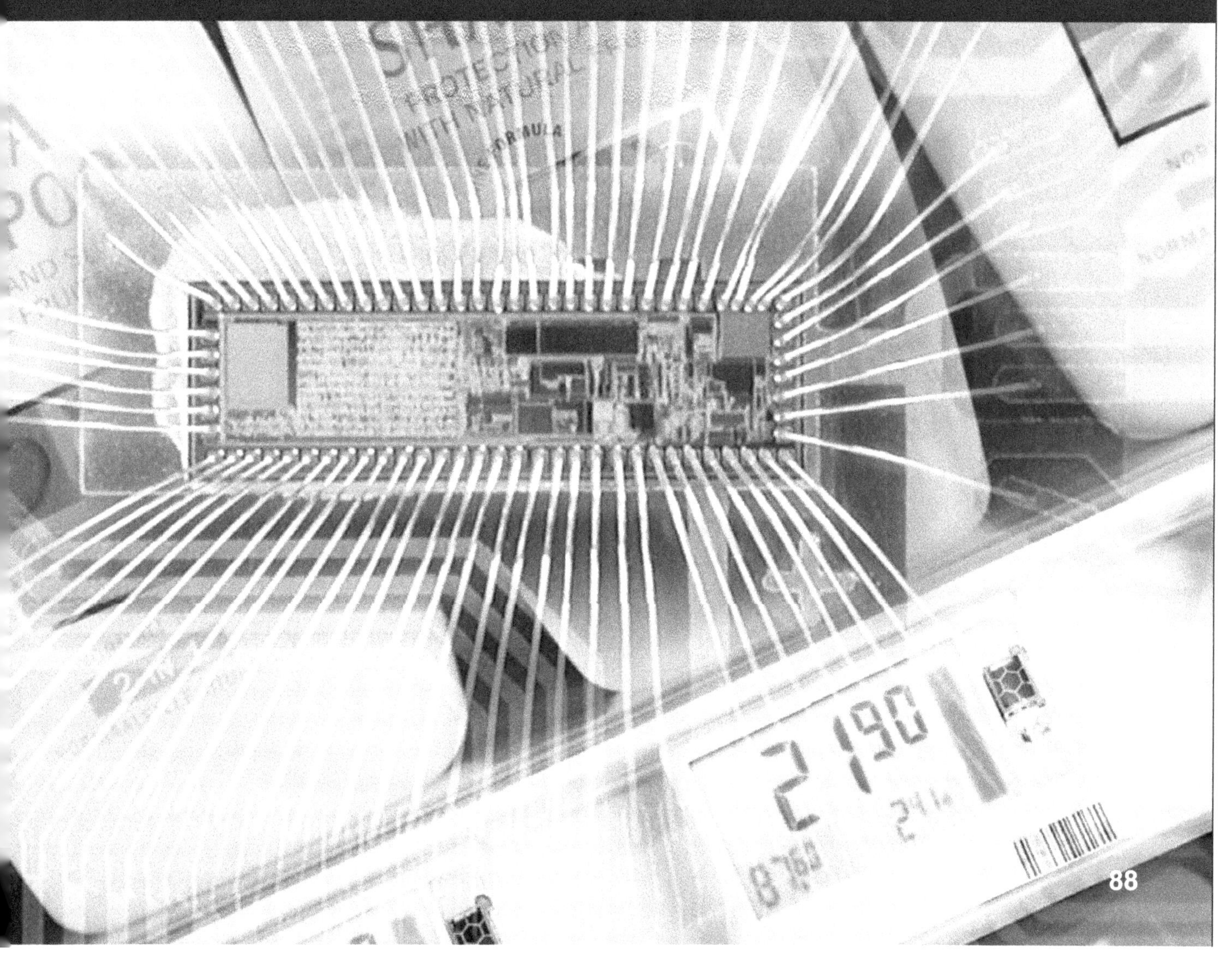

CHAPTER-10

Introduction

Systems that utilise digital signals are referred to as digital electronics. Digital electronics, which are utilised in computers, cell phones, and other consumer goods, are illustrations of Boolean algebra.

Large assemblies of logic gates, which are straightforward electronic representations of Boolean logic operations, are typically used to construct digital electronics or any type of digital circuit.

The terms "digital circuit," "digital system," and "logic" are sometimes used interchangeably when discussing digital circuits by electronic engineers.

10.1: ANALOG AND DIGITAL ELECTRONICS

Let's first define the words in order to clarify the "analogue vs. digital" argument. In the context of digital electronics, there are two possible ways to transmit an electrical signal from point A to point B. *Since they are continuous, analogue signals can have any value.*

Binary numbers are used to encode values in digital signals. A binary number can be conveyed as electronic on/off pulses since it is entirely made up of 0s and 1s (on = 1, off = 0). These pulses are processed as soon as they are received. Physical quantities with discrete variations make create a digital signal.

Each of these signal types communicate information via electrical voltages, and each has benefits and drawbacks. Throughout the audio and recording sector, one of the most contentious analog/digital debates occurs. In this situation, analogue systems can provide a reliable electronic representation of a complicated waveform, to put it a little more simplistically. But because the electrical signal must be amplified, "noise" can be injected along the signal path (fig.10.1). The circuitry's inevitable electron activity is what is causing this cacophony. However, it is difficult to distinguish between this unwanted noise and the genuine signal. As a result, each level of transmission introduces noise into the signal, which can be heard as a "hiss."

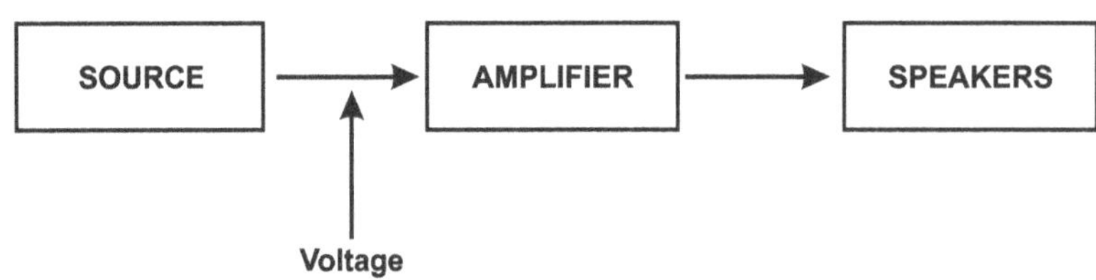

Fig. 10.1 : Analog signal path in a hi-fi system

The system would sample the sound wave at predetermined intervals and communicate the values that, in binary code, correspond to the sound wave, as shown in the digital counterpart in fig.10.2. Without picking up any more noise, the digital representation of the sound wave might then be transferred around or processed within the system.

Even while there is still electron activity (noise), it can be removed at each stage of transmission whenever the digital signal is repeated.

The fig.10.2 system's biggest drawback, aside from its complexity, is that it cannot handle a continuous range of values.

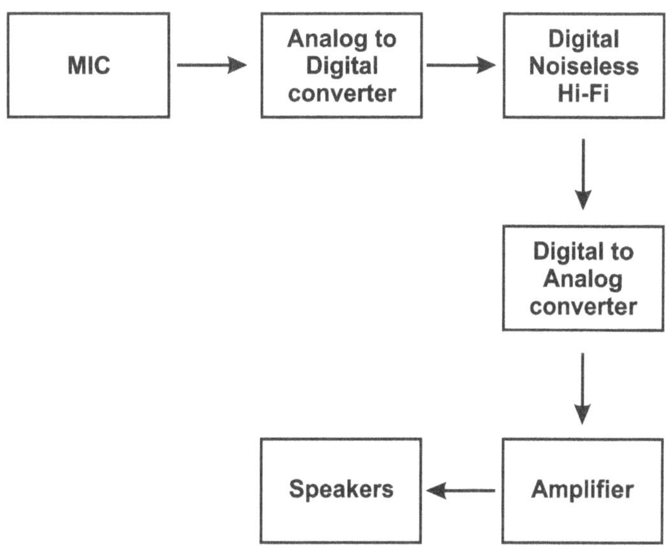

Fig.10.2 : Signal path in a digital/analog hi-fi system

As the sound wave is just digital representations of waves cannot accurately depict them even when they are studied or sampled at certain periods in time.

So, it is easy to summarise the main distinction between analogue and digital values as follows:

Continuous = analogue

Discrete = digital (step by step)

Digital Techniques: Benefits and Drawbacks

Benefits: When compared to analogue circuits, the typical benefits of digital circuits include:

10.2 Benefits of Digital Electronics

a) Digital systems are simple to control with software and work well with computers. A digital system can frequently be expanded with new functionality without requiring hardware changes. By updating the product's software, this may frequently be done outside of the plant. So, once the product is in the hands of the client, any design flaws can be fixed.

b) Digital systems may make it simpler to store information than analogue ones. Data may be stored and retrieved without being harmed because to the noise-immunity of digital systems. The information stored in an analogue system is weakened by noise caused by ageing and wear. Information can be recovered completely in a digital system as long as the overall noise is below a specific threshold.

10.3 DRAWBACKS OF DIGITAL ELECTRONICS

a)   In some circumstances, digital circuits require more power to complete the same functions than analogue circuits, which results in greater heat being produced. This might restrict the usage of digital systems in battery- or portable-powered devices.

b)   For instance, low-power analogue front-ends are frequently used in battery-operated cellular phones to amplify and tune in radio signals from the base station. A base station can use power-hungry but highly adaptable software radios since it is powered by the grid. These base stations are simple to update to handle the signals required by the newest cellular standards.Digital circuits can occasionally cost more, especially when purchased in small quantities.

c)   Signals from the sensed world are analogue quantities, just as the sensed world itself. For instance, analogous phenomena include light, temperature, sound, electrical conductivity, and electric and magnetic fields. The most practical digital systems must convert between discrete digital signals and continuous analogue signals. This results in quantization mistakes.

10.4 STORAGE UNIT OF DIGITAL DATA

A component of the computer system, the storage unit is used to store the data and processing-related instructions. Binary is how digital data is stored. The term "bit" refers to a single binary digit, either a 1 or a 0. A byte is a collection of eight bits, which can be combined in 256 different ways. The fundamental unit of data storage that all other units are often built upon is the byte. A nibble, as it is often called, is a string of four bits that makes up half of a byte.

The term "data storage" refers to all methods and devices used to store digital information on magnetic, optical, or silicon-based media. Offices, data centares, edge settings, remote locations, and private residences all need storage. Storage is an essential component of portable devices like smart phones and tablets. Both individuals and businesses utilize storage as a means to protect information, from private photographs to data that is vital to their operations.

Base-10 decimal and base-2 binary systems are used by computer, storage, and network systems, respectively, to measure storage amounts.

Digital Data Units :

Bit, Nibble, Byte, KB, MB, GB, TB, PB, EB, ZB, YB

1Bit= 0 or 1 , 1 Nibble=4 bits,

1Byte = 8 bits (0000 0000) or (1111 1111) =2^8= 256 Locations

1KB=Kilo Byte =10^3 Bytes = 1024 Bytes 10^3

1MB = 1 Mega Byte =10^3KB =1024 KB = 1,048,576 Bytes 10^6

1GB = 1 Giga Byte =10^3MB =1024 MB = 1,048,576 KB=1,073,741,824 Bytes 10^9

1TB=1 Tera Byte=10^3 GB=1024 GB=1,048,576 MB=1,073,741,824KB= 1,099,511,627,776 Bytes 10^{12}

1PB=1Penta Byte=10^3 TB=1024 TB=1,048,576 GB=1,073,741,824 MB= 1,099,511,627,776 KB= 1,125,899,906,842,624 Bytes 10^{15}

1EB=1Exa Byte=10^3 PB=1024 PB = 1,048,576 TB= 1,073,741,824GB=1,099,511,627,776 MB= 1,125,899,906,842,624 KB=1,152,921,504,606,846,976Bytes 10^{18}

1ZB=1Zetta Byte=10^3 EB=1024 EB=1,048,576 PB= 1,073,741,824TB=1,099,511,627,776 GB=1,125,899,906,842,624 MB=1,152,921,504,606,846,976 KB=1,180,591,620,717,411,303,424 Bytes 10^{21}

1YB = 1 Yotta Byte = 10^3 ZB =1024 ZB = 1,048,576 EB= 1,073,741,824PB= 1,099,511,627,776 TB= 1,125,899,906,842,624 GB=1,152,921,504,606,846,976MB= 1,180,591,620,717,411,303,424 KB=1,208,925,819,614,629,174,706,176 Bytes 10^{24}

10.4 THE DIGITAL NUMBER

Digital technology employs a variety of number systems. The decimal, binary, octal, and hexadecimal systems are the most widely used. We are most familiar with the decimal system because it is a tool we use frequently. We can comprehend the other systems better if we look at some of its properties.

10.5 THE DECIMAL NUMBER

Numeral System Ten numbers or symbols make up the decimal system. These 10 symbols, which are 0, 1, 2, 3, 4, 5, 6, 7, and 8, allow us to express any quantity by utilising them as the digits of a number. Because it has 10 digits, the decimal system is sometimes known as the base-10 system.

MSB - Most Significant Bit 1000

LSB - Least Significant Bit 0.0001

$10^3 = 1000$ $10^{-1} = .1$

$10^2 = 100$ $10^{-2} = .01$

$10^1 = 10$ $10^{-3} = .001$

$10^0 = 1$ $10^{-4} = .0001$

10.6 Binary Number

There are just two potential digit values or symbols in the binary system: 0 and 1. Any quantity that can be expressed in decimal or another number system can be represented using this base-2 system.

$2^3 = 8$ $2^{-0} = 0$

$2^2 = 4$ $2^{-1} = 1/2$

$2^1 = 2$ $2^{-2} = 1/4$

$2^0 = 1$ $2^{-3} = 1/8$

10.7 Counting of Binary Number

In the table in figure 10.3, the binary counting sequence is displayed.

$2^3=8$	$2^2=4$	$2^1=2$	$2^0=1$	Decimal Equivalent
0	0	0	0	0
0	0	0	1	1
0	0	1	0	2
0	0	1	1	3
0	1	0	0	4
0	1	0	1	5
0	1	1	0	6
0	1	1	1	7
1	0	0	0	8
1	0	0	1	9
1	0	1	0	10
1	0	1	1	11
1	1	0	0	12
1	1	0	1	13
1	1	1	0	14
1	1	1	1	15

Fig.10.3 : Signal path in a digital/analog hi-fi system

10.8 Representing Binary Quantities

The information being processed in digital systems is typically displayed in binary form. Any gadget with two different operating states or circumstances can represent binary quantities. A switch, for instance, can only be opened or closed. We have shown that we can represent any binary number arbitrarily by using a series of switches.

Binary 1: Any voltage from 2 volts to 5 volts

Any voltage between 0 and 0.8 volts is binary 0.

Never used Voltage between 0.8V and 2V could result in a digital circuit being faulty.

Another clear distinction between digital and analogue systems is this.

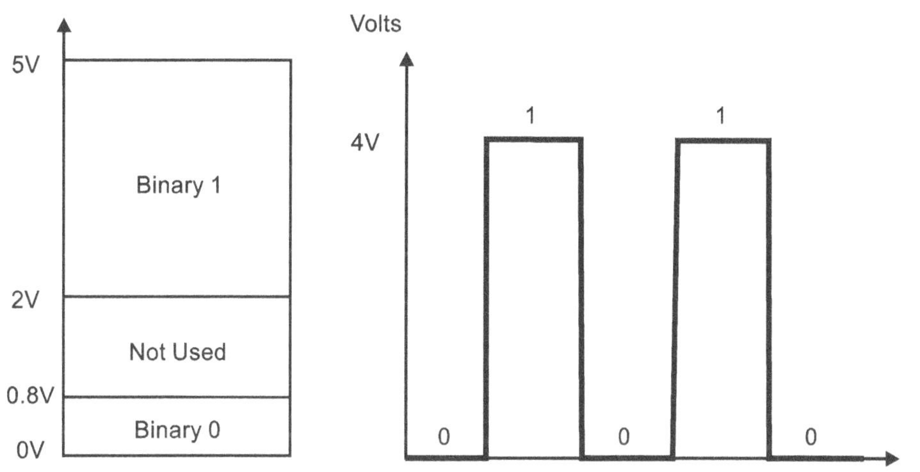

CHAPTER-11

Logic Gates And Boolean Algebra

11.1 LOGIC GATES

A logic gate, a simple electrical circuit, is frequently used to build a digital circuit. A particular boolean logic operation is represented by each logic gate. Electrically controlled switches are arranged in a logic gate.

A logic gate's output is a voltage or electrical current that can control further logic gates when used in series.In order to decrease their size, power consumption, cost, and reliability, logic gates frequently employ the fewest possible transistors.

The cheapest approach to manufacture logic gates in large quantities is using integrated circuits. Engineers typically use software for electronic design automation to create integrated circuits.

The basic components of digital circuitry are logic gates, sometimes known as "gates." They work by "opening" or "closing," as their name suggests, to accept or reject the flow of digital information. Simple logical operations on boolean (Boolean algebra) variables, i.e., variables that can only have one of two states (0/1, low/high, false/true), are implemented electronically by Gates. Electrically speaking, any voltage between 0 and 0,7 volts and between 2,5 and 5, volts, respectively, represent the logic states 0 and 1, for the TTL (transistor-transistor-logic) family of digital devices. The accepted electronic symbols for various gates, along with their related "truth tables" and symbolic logical statements, are shown in the following picture as fig. 11.1. (Y, A, B,...) are all boolean variables.

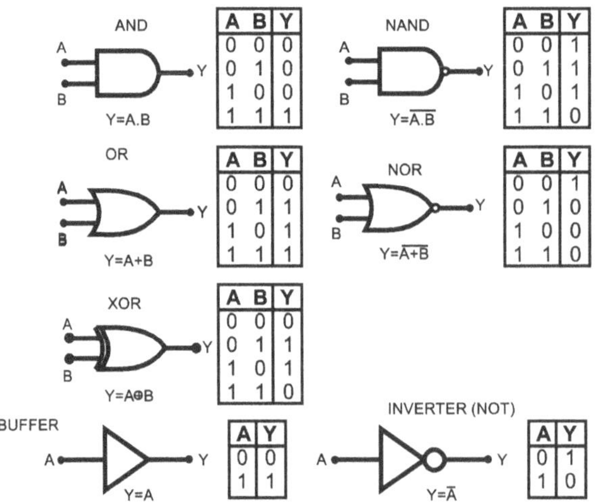

Fig.11.1 Logic gates

Figure 11.1 depicts many logic gates. You can clarify the concepts using truth tables and symbols, as well as the following points.

AND GATE

the equation Y = A. B can be taken into account when "Y equals A AND B."

The multiplication sign denotes the AND operation, just like it does for standard 1:0:1 multiplication..

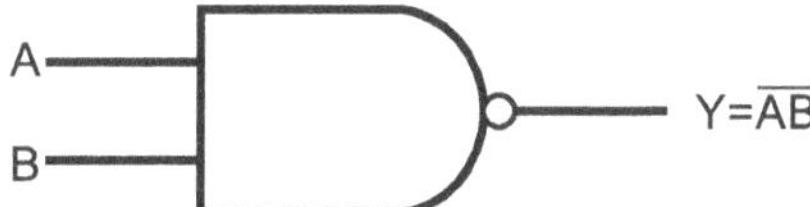

When all of the input variables are 1, which happens just once, the AND operation yields a result of 1.

NAND GATE

When one or more inputs are 0 NAND, the output is always 0. With the exception of a tiny circle on the output, the gate NAND symbol is identical to the AND gate symbol. The inversion operation is represented by this tiny circle. As a result, Y = AB is the two input NAND gate's output expression.

OR GATE

It says "Y equals A OR B" when Y = A + B is used. In contrast to regular addition, the + symbol denotes the OR operation. When either of the input variables is 1, OR produces a result of 1.

Only when each and every one of the input variables is 0, does the OR operation return a result of 0.

NOR GATE

NOR Digital circuitry frequently uses the gate types NOR and NAND. It is quite simple to explain these gates using boolean algebra because they integrate the fundamental operations AND, OR, and NOT. The only difference between the NOR and OR gate symbols is that the output of the NOR gate contains a little circle. This tiny circle is an illustration of how an inversion works.

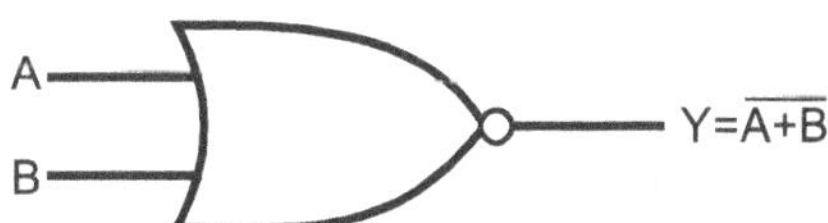

XOR GATE

The XOR function, also known as the Exclusive-OR function, can be orally defined as "either A or B, but not both." There are a variety of more thorough and accurate ways to state this in the context of digital logic.

We won't discuss tools like Truth tables and visual representations here. We'll continue to use the clearer verbiage "A and NOT B, or NOT A and B."

NOT Gate

Because just one input variable can be used, the NOT operation differs from the OR and AND procedures. As an illustration, the result x can be written as follows if the variable A is subjected to the NOT operation:

In the formula $Y = \overline{A}$, the NOT operation is denoted by the prime (-). It reads as Y equals NOT A in this expression.

The reciprocal of A is Y.

A's complement, Y, is equal.

These all show that the logic value of $Y = A'$ is the opposite of the logic value of A and are all accepted terms.

It is common to refer to the NOT operation by other names, such as complementation or inversion.

10.2 CONCEPT OF BOOLEAN ALGEBRA

The mathematics of logical expressions, first introduced by George Boole in 1854 and nowadays known as Boolean Algebra, is a method to reduce logical expressions. Any logical expression can be subjected to Boolean Algebra's clear-cut rules, which are basic and straightforward. An easy way to determine whether the reduction was successful is to quickly test the reduced phrase using a Truth Table.

Finding ways to make digital circuits as simple as feasible is one of the main demands when working with them. This repeatedly necessitates the transformation of complicated logical formulations into phrases that are simpler but nevertheless provide the same outcomes in all conceivable circumstances. The simplified expression can then be implemented using a smaller, simpler circuit, which saves on the cost of the extra gates, lowers the required number of gates, and lowers the power and area requirements of those gates.

Boolean Algebra has the following rules:

OR Operations (+)		AND Operations (·)	
$0+0 = 0$	$X+0 = X$	$0 \cdot 0 = 0$	$X \cdot 0 = 0$
$1+0 = 1$	$X+1 = 1$	$1 \cdot 0 = 0$	$X \cdot 1 = X$
$0+1 = 1$	$X+X = X$	$0 \cdot 1 = 0$	$X \cdot X = X$
$1+1 = 1$	$X+\overline{X} = 1$	$1 \cdot 1 = 1$	$\overline{X} \cdot X = 0$

Boolean Algebra has the following rules:

Associative Law :

$(X+Y)+Z=X+(Y+Z)=X+Y+Z$ (OR) , $(X.Y).Z=X.(Y.Z)=X.Y.Z$ (AND)

Distributive Law :

$X+(Y.Z)=(X+Y).(X+Z)$ (AND), $X(Y+Z)=XY+XZ$ (OR)

Commutative Law :

$X+Y=Y+X$ (OR), $X.Y=Y.X$ (AND)

Identity Law :

$X+0=X$ (OR), $X.1=X$ (AND)

Redundancy Law :

$X.(\overline{X}+Y)=X.Y$, $X+\overline{X}Y=X+Y$

11.3 DE MORGAN'S THEOREM

In order to simplify formulas where a product or the sum of variables is inverted, DeMorgan's theorems are quite helpful. They are as follows:

$\overline{A.B}=\overline{A}+\overline{B}$ (NAND)

$\overline{A+B}=\overline{A}.\overline{B}$ (NOR)

According to Theorem (NAND), when two variables' AND products are inverted, doing so is equivalent to inverting each variable separately before ORing them.

According to Theorem (NOR), inverting the OR sum of two variables has the same effect as independently inverting each variable before ANDing the resulting inverted variables.

11.4 THE LOGIC EQUIVALENT CIRCUIT

More complexity can be achieved by combining logic gates. A combination of them can be used to substitute one kind of gate for another.

As shown in the truth table on the right, for instance, we may combine a NOT gate and an AND gate to create an output Q that is true only when inputs A and B are true.

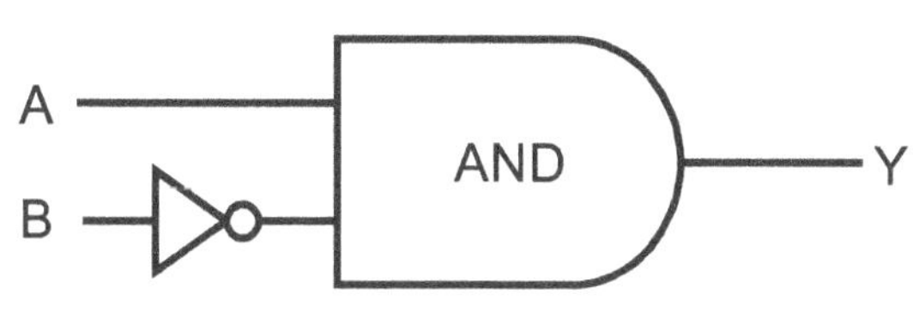

Input A	Inuput B	Output Y
0	0	0
0	1	0
1	0	1
1	1	0

Fig.11.2 : Y = A AND NOT B

11.5 REPLACING ONE TYPE OF GATE FOR ANOTHER

Altering the type of a gate Logic gates are available on integrated circuits (ICs), which typically contain many gates of the same kind, such as four 2-input NAND gates or three 3-input NAND gates. If only a few gates are needed, this can be inefficient unless they are all the same type. You can decrease the number of gate inputs or swap out one type of gate for another to prevent employing too many ICs.

11.6 MINIMISING THE NUMBER OF INPUTS

By coupling two (or more) inputs together, the number of inputs to a gate can be minimised. An AND gate with three inputs that can also accept two inputs is shown in the diagram.

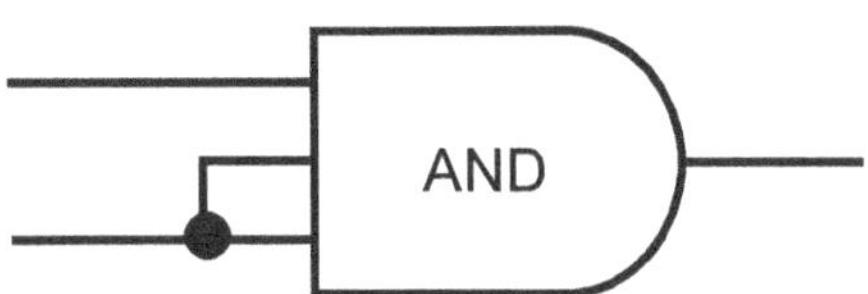

11.6: HOW TO CONVERT A NAND OR NOR GATE INTO A NOT GATE

A NOT gate is produced when a NAND or NOR gate has only one input. With a 2-input NAND gate, this is depicted in the diagram.

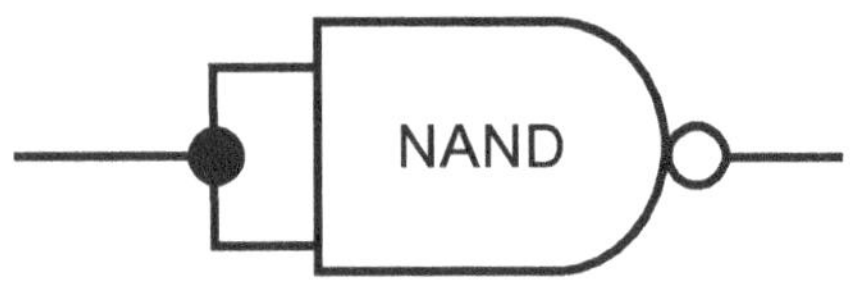

11.7 NAND OR NOR GATES CAN BE COMBINED TO CREATE ANY GATE.

NAND or NOR gates can be coupled to make any kind of gate, in addition to creating a NOT gate. This makes it possible to construct a circuit using just one kind of gate—either a NAND or a NOR gate. For instance, a NAND gate follows a NOT gate after an AND gate (to undo the inverting function). Because they lack the inverting (NOT) function, AND and OR gates cannot be combined to form other gates.

Three steps must be taken in order to switch an OR gate to an AND gate:

Invert each input (NOT). Change the gate type from OR to AND or FROM AND TO OR. Invert the output (NOT). For instance, NOTed inputs fed into a NAND (AND + NOT) gate can be used to construct an OR gate. OR gate can be built from NOTed inputs fed into a NAND (AND + NOT) gate.

The NAND gate's equivalent circuit: NOT, AND, OR, and NOR equivalents of the NAND gate are shown in the table below.

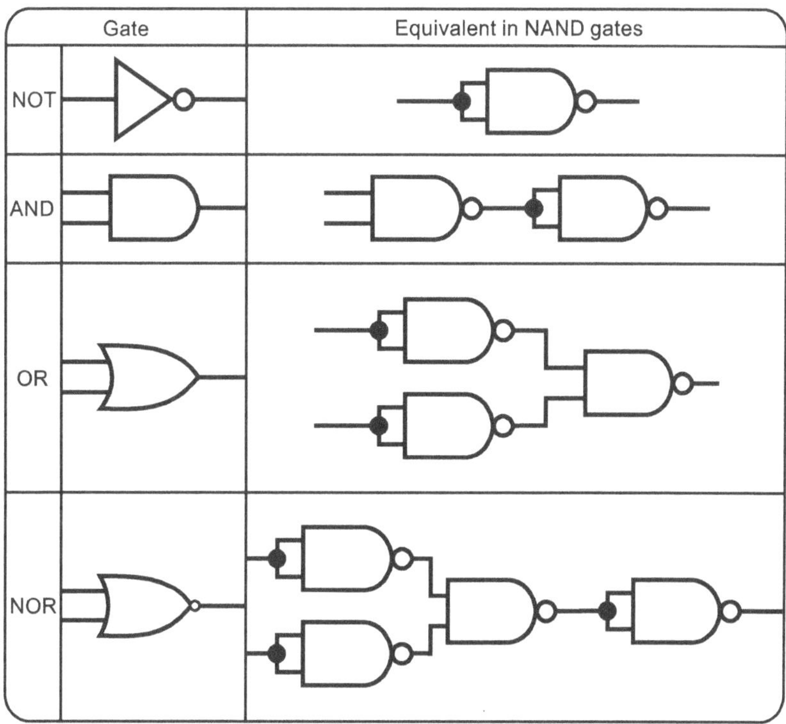

11.9   GATE SUBSTITUTION IN A SIMPLE LOGIC SYSTEM

Three distinct gates—NOR, AND, and OR—

are present in the original scheme.

Three ics are needed (one for each type of gate).

The diagram below illustrates how to redesign this system using NAND gates solely by substituting each gate with a NAND equivalent.

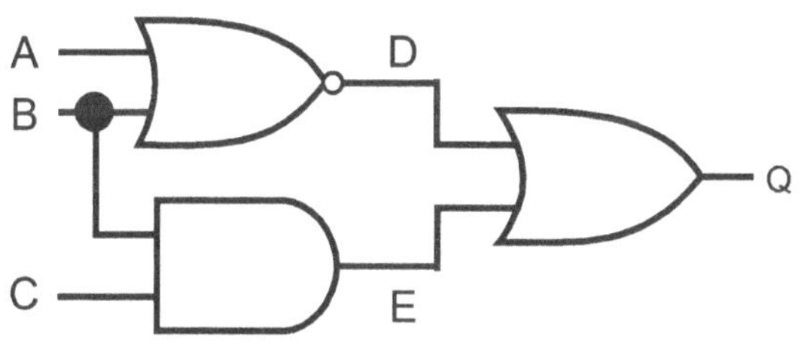

Here is an illustration of the completed system. It contains five NAND gates and requires two ICs, each with four gates. Compared to the prior approach, which asked for three Ics (one for each type of gate), this is an improvement.

NAND (or NOR) gates are not always replaced with NOR gates, but when they are, as in this example, the number of gates increasesis typically limited to one or two gates. The main advantage comes from employing just one kind of gate, which reduces the number of ICs needed.

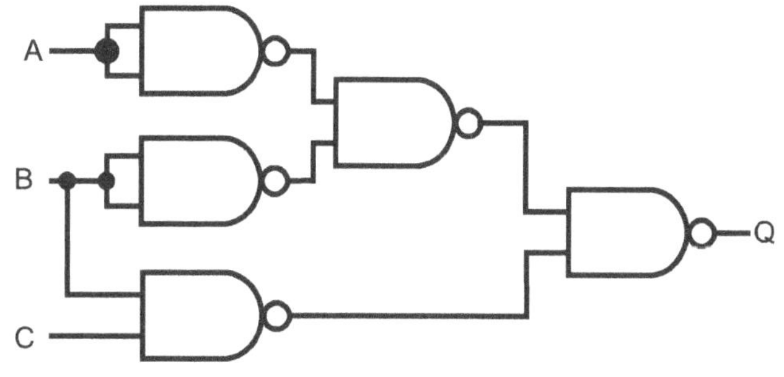

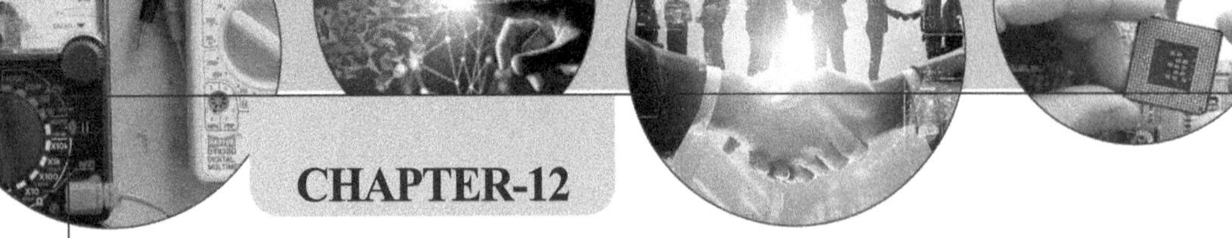

CHAPTER-12

Binary Number System

The most crucial number system in digital systems is the binary one, but there are a few others that are equally significant. Since it is always used to represent amounts other than those in a digital system, the decimal system is significant. Hence, there will be instances where decimal numbers must first be transformed to binary values before being input into the digital system.

Digital electronics and binary numbers have many similarities. In digital electronics, a "1" indicates the presence of current or electricity, while a "0" indicates the absence of such. Pulses of current are used by a computer's components to interact with one another (1s and 0s).

We are all aware of how quickly computers can solve difficult problems and do advanced maths. The BINARY SYSTEM is a method of calculation that uses just 1s and 0s. Despite the fact that computers can only handle 1s and 0s, there comes a time when the 1s and 0s must be transformed into the familiar decimal numbers we are used to.

When attempting maths, we frequently employ the DECIMAL SYSTEM. This approach works with numbers that we frequently use, such as 1, 2, 3, 5, 6, 7, 8, 9, 10s, 100s, and 1000s, etc. Since there are only two numbers in the BINARY system—1s and 0s—you might be wondering how it is possible to count past one.

Two other number systems, in addition to binary and decimal, are widely used in digital systems. Both the octal (base-8) and hexadecimal (base-16) number systems are employed to represent huge binary systems in an effective manner.

12.1 BINARY-TO-DECIMAL CONVERSION

Just adding up the weights of the different spots in the binary number that contain a 1 will allow you to convert any binary number to its decimal counterpart.

Binary Decimal

$$(1\ 1\ 0\ 1\ 1)_2 = 2^4 \text{x}1 + 2^3 \text{x}1 + 2^2 \text{x}0 + 2^1 \text{x}1 + 2^0 \text{x}1 = 16 + 8 + 0 + 2 + 1 = (27)_{10}$$

As you can see, the procedure is to add the weights (i.e., powers of 2) for each bit location that contains a 1 after finding the weights (i.e., powers of 2).

12.2: DECIMAL -TO- BINARY CONVERSION

Here, each decimal digit is individually transformed to binary. $7 = (0111)_2$ and $9 = (1001)_2$, $27 = (11011)_2$ then $79 = 01111001$. We can grasp this conversion with the help of the table in Fig. 3.1.

When decimal numeral displays are utilized, for instance, it is fairly convenient.

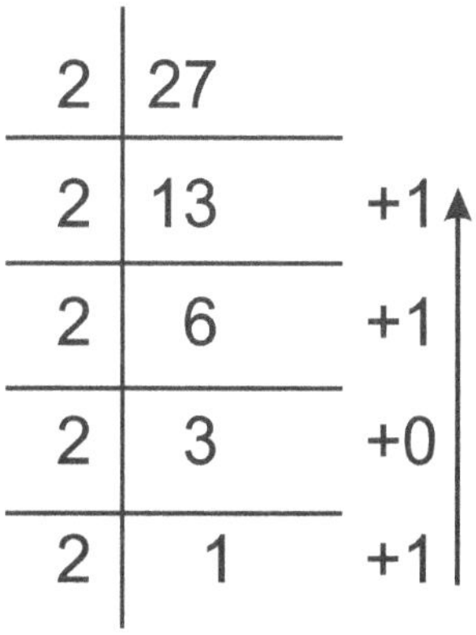

$$\begin{array}{r|l}
2 & 27 \\
\hline
2 & 13 \quad +1 \\
\hline
2 & 6 \quad +1 \\
\hline
2 & 3 \quad +0 \\
\hline
2 & 1 \quad +1 \\
\end{array}$$

fig.3.1 This method uses repeated division by 2. Eg. convert $(27)_{10}$ to binary

$(27)_{10} = (11011)_2$

We now have a better understanding of the binary to decimal example from earlier.

12.3 OCTAL NUMBER

Since the octal number system has an eight-digit base, there are eight potential digits: 0, 1, 2, 3, 4, 5, 6, and 7.

12.4 CONVERSION OF OCTAL TO DECIMAL

8^3	8^2	8^1	8^0		8^{-1}	8^{-2}	8^{-3}
=512	=64	=8	=1	.	=1/8	=1/64	=1/512
Most **S**ignificant **D**igit				Octal point			**L**east **S**ignificant **D**igit

12.5 DECIMAL AND OCTA-DECIMAL CONVERSION

Octal Digit	0	1	2	3	4	5	6	7
Binary Equivalent	000	001	010	011	100	101	110	111

is technique utilises eight times divided. for instance, binary and octal conversion of 177:

$$177_{10} = 261_8$$

Now we can convert octal to binary using the table below.

$$261_8 = (010110001)_2$$

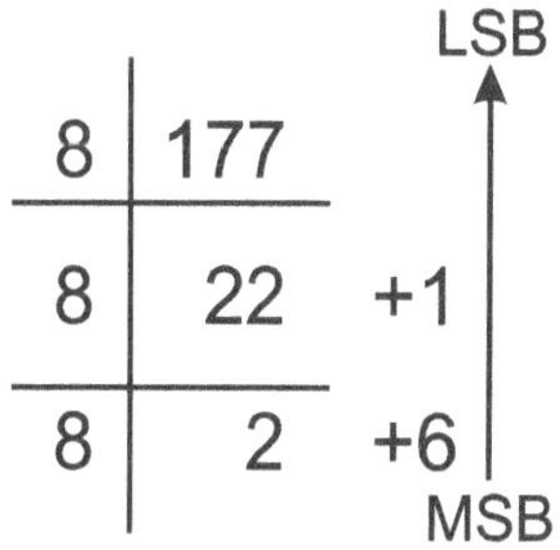

$$\begin{array}{r|l}
8 & 177 \\
\hline
8 & 22 \quad +1 \\
\hline
8 & 2 \quad +6 \\
\end{array}$$

LSB

MSB

12.6 DECIMAL AND HEXADECIMAL CONVERSION

Representing blocks of four bits with a single number is frequently highly beneficial. Hexadecimal, sometimes known as base-16, is a convention that uses one digit to represent the numerals 0, 1, 2, :::, and 15. It uses letters A–F after following decimal for 0–9.

Decimal	Binary	Hex
0	0000	0
1	0001	1
2	0010	2
3	0011	3
4	0100	4
5	0101	5
6	0110	6
7	0111	7
8	1000	8
9	1001	9
10	1010	A
11	1011	B
12	1100	C
13	1101	D
14	1110	E
15	1111	F

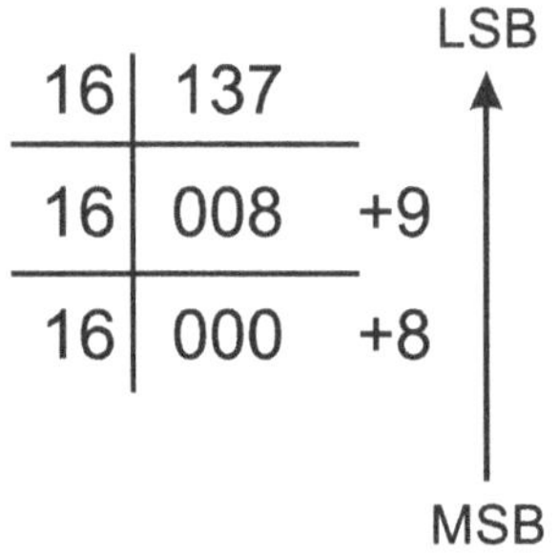

Decimal to Hexa
$(137)_{10} = (89)_{16}$

Hexa to Decimal
$16^1 \times 8 + 16^0 \times 9 = 137$

The capability of using logical functions to carry out mathematical operations is a crucial necessity for digital computers. The fundamental operation here is addition; if we can add two binary integers, we can remove them just as easily or, for something a little fancier, conduct multiplication and division. In such case, how do we add two binary numbers?

First, let's add two binary bits. There are only four potential input combinations because each bit can only have one of the two possible values, 0 or 1.

Binary addition produces the answer "10," with a "1" in the "twos place" and a "0" in the "ones place," rather than "2" (a symbol that does not exist in the binary number system). You would recite, "One and one are two; write down the zero, carry the one," if you wrote this addition vertically:

12.7 RULES FOR ADDITION ARE:

a) $0 + 0 = 0$

b) $0 + 1 = 1$

c) $1 + 0 = 1$

d) $1 + 1 = 10$

e) $10 + 1 = 11$

Consider this:

Let's look at a binary addition problem:

$$11100$$
$$+\ 11010$$
$$\overline{\qquad ?\qquad}$$

The least-significant column should be added first to obtain

$$11100$$
$$+\ 11010$$
$$\overline{\qquad 0\qquad}$$

Afterward, include the bits in the second column.

$$11100$$
$$+\ 11010$$
$$\overline{\qquad 10\qquad}$$

The third column displays.

$$11100$$
$$+\ 11010$$
$$\overline{\qquad 110\qquad}$$

Carry appears on the fourth column.

Carry
$$\longrightarrow 1$$
$$11100$$
$$+\ 11010$$
$$\overline{\qquad 110\qquad}$$

Take note of the carry, which happens since $1+1=10$, into the last column. You enter the 0 and carry the 1 to the column above it, just like in decimal addition.

In the final column, Carry is provided.

Carry
$$\longrightarrow 1$$
$$11100$$
$$+\ 11010$$
$$\overline{110110}$$

The last column's formula is $1+1+1=10+1=1$.

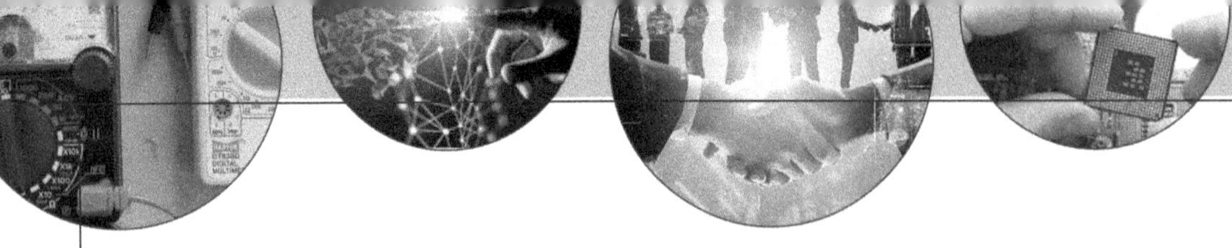

12.8   SUBTRACTION OF BINARY NUMBERS

The guidelines for binary subtraction are as follows:

a) $0 - 0 = 0$

b) $1 - 0 = 1$

c) $1 - 1 = 0$

d) $10 - 1 = 1$

Similar to how you would with decimal numbers, subtract larger binary integers column by column. You occasionally have to borrow from the column above, thus this. As an illustration:

$$\begin{array}{r} 1100 \\ -\ 1010 \\ \hline ? \end{array} \qquad \begin{array}{r} 1100 \\ -\ 1010 \\ \hline 0 \end{array}$$

To obtain, take the LSBs out.

$$\begin{array}{r} \longrightarrow \quad 1 \\ 1100 \\ -\ 1010 \\ \hline 10 \end{array}$$

Borrow from the next-higher column to get the bits you need to subtract the bits from the second column.

$$\text{Borrow} \quad \begin{array}{r} 1100 \\ -\ 1010 \\ \hline 0010 \end{array}$$

Subtraction to the second column over from the right: $10 - 1 = 1$, to get

The remaining columns are then subtracted:

Binary subtraction is not more challenging than decimal subtraction once you get the hang of it.

12.9   FUNCTION OF A HALF ADDER

A "half adder" is the name of this electronic component. The term "half adder" describes the fact that, despite being able to produce a signal to signify a carry to the next highest order bit, this configuration is unable to take a carry from a lower-order adder.

The Sum is an Exclusive-OR function, while the Carry output is a straightforward AND function. As a result, we can combine these two bits using two gates. The circuit that results is displayed below.

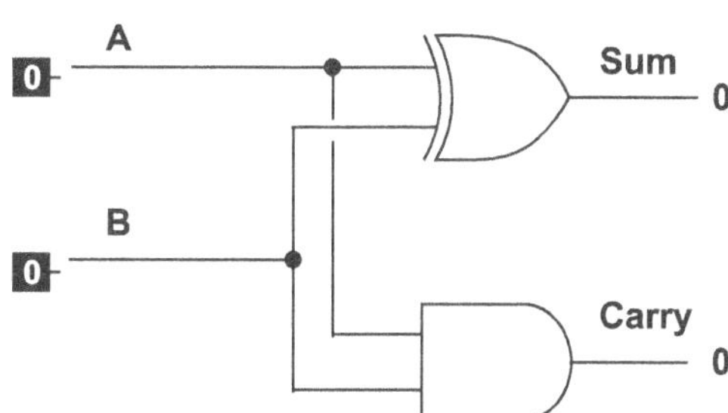

<table>
<tr><th colspan="2">INPUTS</th><th colspan="2">OUTPUTS</th></tr>
<tr><th>A</th><th>B</th><th>CARRY</th><th>SUM</th></tr>
<tr><td>0</td><td>0</td><td>0</td><td>0</td></tr>
<tr><td>0</td><td>1</td><td>0</td><td>1</td></tr>
<tr><td>1</td><td>0</td><td>0</td><td>1</td></tr>
<tr><td>1</td><td>1</td><td>1</td><td>0</td></tr>
</table>

fig.12.1 : Half Adder Circuit and Truth Table

Each pair of bits must be able to generate an output carry and be able to detect and include a carry from the next lower order of magnitude if they are to function properly.

If there is a carry from one column to the next, the carry must be included in the following column. This is the same condition as adding decimal numbers. For the same reason, we must also perform the same action with binary numbers. As a result of just performing half of the work, the circuit on the left is referred to as a "half adder." We require a circuit that can do the entire task.

12.10   FUNCTION OF A FULL ADDER

AThe inputs of a "full adder" are three. Together with the two addends, there is a third input called "carry in" that adds the bit carried over from the preceding column.

We will refer to them as CIN and COUT because we will have both an input carry and an output carry. We'll also use S to denote the output of the final Sum calculation. In fig. 12.2, the resulting truth table is displayed.

A	B	C_{IN}	C_{OUT}	S
0	0	0	0	0
0	0	1	0	1
0	1	0	0	1
0	1	1	1	0
1	0	0	0	1
1	0	1	1	0
1	1	0	1	0
1	1	1	1	1

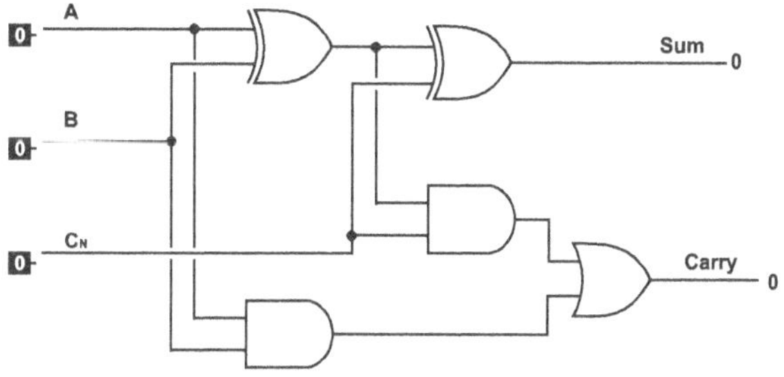

fig.12.2 : Full Adder Circuit and Truth Table

The first will combine A and B to form a partial Sum, and the second will increase that Sum by CIN to form the complete S output. There will be an output carry if either half-adder generates a carry. As a result, COUT will be an OR function of the Carry outputs from the half-adder. As a result, COUT will be an OR function of the Carry outputs from the half-adder. This is a diagram of the resulting full adder circuit.

Depending on the value of A, it appears that COUT may either be an AND or an OR function, and S may either be an XOR or an XNOR. The S output, on the other hand, is actually an XOR between the A input and the half-adder SUM output with B and CIN inputs, as can be seen by paying closer attention. Moreover, if any two or all three of the inputs are logic 1, the output carry will be true. Equations can be used to express it as follows:

$$Sum = \overline{A}\overline{B}C + \overline{A}B\overline{C} + A\overline{B}\overline{C} + ABC$$
$$Carry = \overline{A}BC + A\overline{B}C + AB\overline{C} + ABC$$

A distinct symbol, illustrated below, is used to represent a one-bit full adder because the full adder circuit depicted in fig.12.2 is actually too complex to be utilised in bigger logic diagrams.

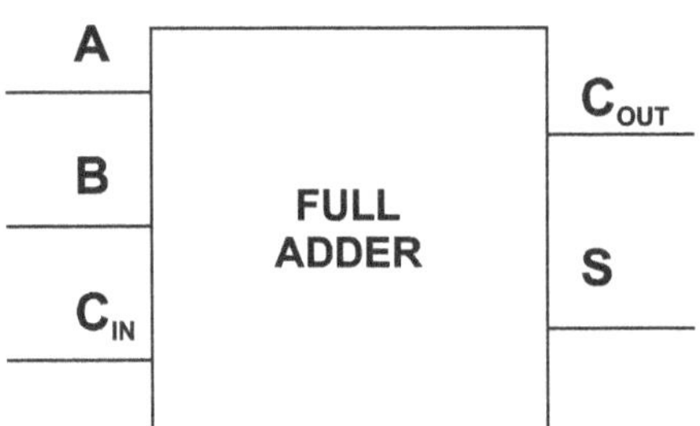

A full adder must be reserved in order to add several bits simultaneously, just like a computer would. Hence, four complete adders with cascaded carry lines are required to add two 4-bit numbers to yield a 4-bit sum (perhaps with a carry), as seen to the right. Eight full adders, which can be created by cascading two of these 4-bit blocks, are required to add two 8-bit values. By extension, this method can be used to combine two binary numbers of any size.

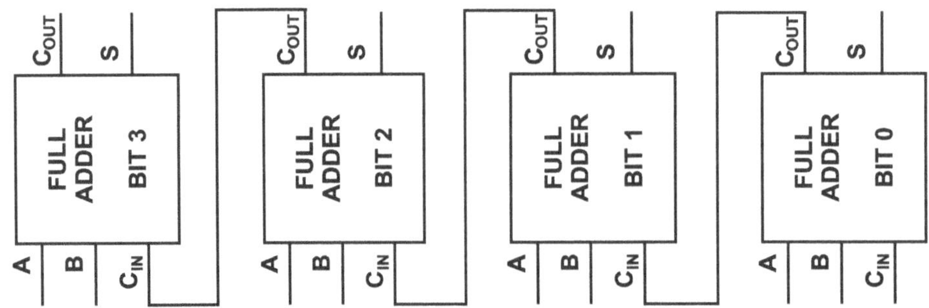

12.11 SUBSTRACTOR AND MINUS NUMBERS

For displaying negative numbers, there are two accepted conventions. A negative number is shown by the MSB using sign magnitude. 0011 = 3 and 1011 = - 3 are examples of 4-bit numbers, for example. Despite being straightforward to understand, this is bad for maths..

Negative numbers are constructed with the concept of the 2's complement to ensure that the sum of a number and its 2's complement is zero. To continue with the 4-bit example, 0101 = 5 and its 2-compliment, 5 = 1011, are given. The result of adding is 10000 = 0 (remember to carry). (The fifth bit is not considered.) Using the 2's complement results in addition and multiplication as expected. Forming the 2's complement can be done in one of two ways:

Then, transform 0 to 1 and 1 to 0, then add 1.

2. To get the desired number, add a number to 2MSB. An illustration of determining the 2's complement of a 4-bit value is - 5 = - 8 + 3 = 1000 + 0011 = 1011.

We have shown how straightforward logic gates are capable of doing binary addition. It stands to reason that a circuit identical to the one in fig. 12.3 might carry out binary subtraction.

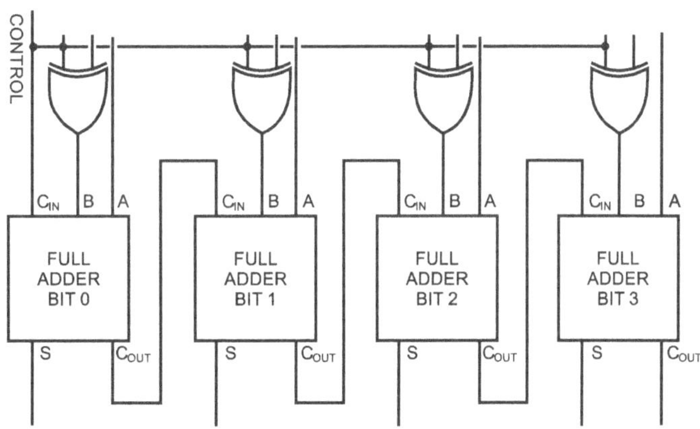

fig. 12.3 : Substracter

We can easily adapt our initial 4-bit adder circuit to an adder/subtractor now that we know how to quickly find the negative of any value. The output of A + B is obtained by maintaining the inputs in tact. The outcome is A - B, though, if we invert B and multiply 1 by the low-order Cin. To decide whether to add or remove on any given instance, we can utilise Exclusive-OR gates, as shown in fig.3.4. The XOR gates will maintain the B input number if the control input is 0, and they will also apply logic 0 as the initial input carry. To add the two numbers, we specifically want this.

On the other hand, if we add a logic 1 to the control input, the XOR gates will invert the B input number to create its one's complement and will also add 1 through the original input carry. B now has the complement of two.

The actual output result will therefore be A - B. (Remember that the output carry is disregarded in two's complement addition. In addition, you can consider it to be an inverted "borrow" bit rather than a carry, making a carry of 1 equal to a borrow of 0. This reasoning also applies to the input carry, which is a logical representation of an input borrow bit of 0.)

12.12 DIGITAL DATA ENCRYPTION

ASCII (American Standard Code for Information Interchange) characters are used for the majority of data exchange. A parity bit is added to 7-bit ASCII codes to produce an 8-bit digital integer. Computer passwords, electronic door locks, and banking equipment all use some type of encryption and ASCII data to maintain security. For the purpose of encrypting ASCII data, the 8-bit pseudo-random number generator is a helpful circuit. The sequence can actually start at any initial value other than the forbidden condition (11111111). Let's say the original value was (01111010), also known as 122 in decimal form, r in HEX, or the ASCII character "z." The PRNG sequence is just offset by this amount, but it continues to repeat after 255 cycles as usual. A representation of an 8-bit PRNG starting at some index (7) and the next six values is shown below in a Boolean array. As you can see, the sequences are identical and predictable after 255 cycles plus this index ($7 + 255 = 262$).

12.13 BCD-BINARY CODED DECIMAL

Not all digital arithmetic is carried out simply directly translating into base-2. Another representation is known as binary coded decimal, or BCD. According to the following table, each decimal digit in BCD is individually encoded in four bits:

Decimal Digit	BCD Representation	Decimal Digit	BCD Representation
0	0000	5	0101
1	0001	6	0110
2	0010	7	0111
3	0011	8	1000
4	0100	9	1001

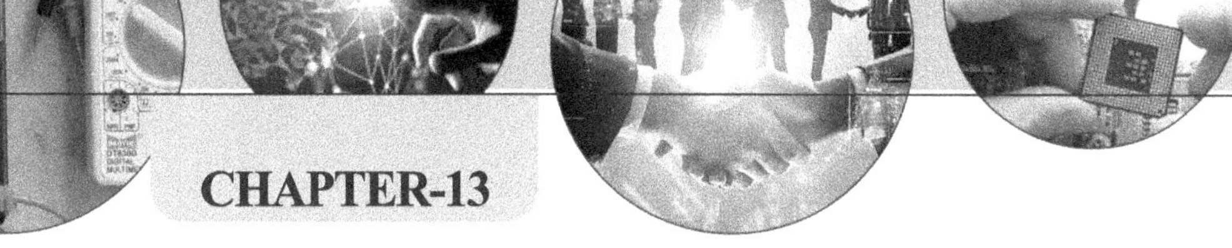

CHAPTER-13

Flip-Flop

The flip-flop, which has two stable states and can therefore be used as one bit of memory in digital circuits, is a type of biteable multi vibrator. The name "flip-flop" now typically refers to non-transparent (clocked or edge-triggered) devices, whilst the more straightforward transparent ones are sometimes referred to as "latches."

A flip-flop is managed by one or more control signals (often one or two) and/or a gate or clock signal. In addition to the expected output, the output frequently contains the complement. Flip-flops are implemented electrically, therefore they logically also need connections for power and ground.

Flip-flops come in two varieties: simple (transparent) and timed. Two cross-coupled inverting elements, such as transistors, NAND, or NOR-gates, can be used to create basic flip-flops, which may also have an enable/disable (gating) mechanism. Devices that are specifically designed to be clocked for synchronous (time-discrete) systems ignore their inputs except when a specialised clock signal changes (this is known as clocking or pulsing). As a result, depending on the values of the input signals at the transition, the flip-flop either changes or maintains its output signal. Flip-flops can adjust their output on either the rising or falling edge of the clock.

13.1 RS FLIP-FLOPS

The simple RS latch, also known as a simple RS flip-flop, is the most basic latch. A pair of cross-coupled NOR (negative OR) logic gates can be used to build it. The output with the letter Q has the stored bit.

A simple NOR latch is shown in Figure 13.1's circuit. The inputs are commonly designated "S" and "R" for "Set" and "Reset," respectively. The NOR inputs in this circuit must normally be logic 0 to prevent overriding the latching action, hence the inputs are not reversed in this design. This is a circuit with a NOR-based latch:

The S and R inputs are typically both low while the storage mode is engaged, and feedback keeps the Q and $\overline{Q}$ outputs—Q being the complement of $\overline{Q}$—in a constant state. The Q output is pushed high if S (Set) is pulsed high while R is held low, while the Q output is forced low if R (Reset) is pulsed high while S is held low. In both cases, the $\overline{Q}$ output remains low even after $\overline{S}$ returns low.

It is a banned condition if both inputs are present at the same time at the logic 1 level. That condition will override the feedback latching operation and force both outputs to a logic 0. In this case, the input that reaches logic 0 first loses control while the other input, which is still at logic 1, determines the latch's final state

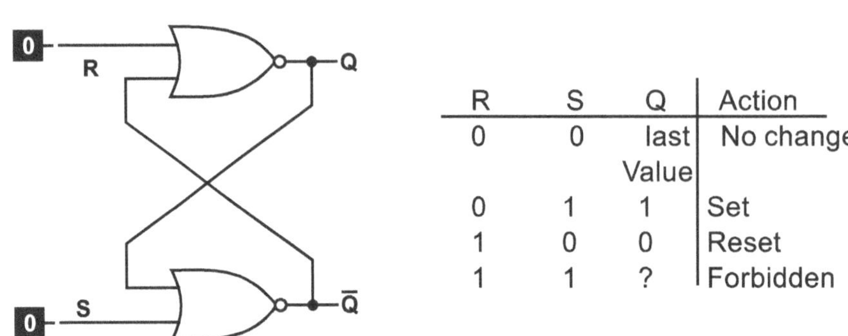

R	S	Q	Action
0	0	last Value	No change
0	1	1	Set
1	0	0	Reset
1	1	?	Forbidden

fig.13.1 : RS-Flip-Flop and Truth Table

13.2 Flip-Flop RS Clocked

Two objectives are achieved when two NAND gates are added to the RS latch's input circuits: regular inputs as opposed to inverted inputs, and a third input that is shared by both gates and can be used to synchronise this circuit with similar ones.

The basic latch you looked at on the previous page operates quite similarly to the clocked RS latch circuit. To modify the latch's state, the S and R inputs, which are typically at logic 0, must be changed to logic 1. However, a new factor has been added with the third input. Since a clock circuit of some kind is generally used to synchronise a number of these latch circuits with one another, this input is frequently called C or CLK. Only when the CLK input is a logic 1 can the output change states. The S and R inputs have no impact when CLK is a logic 0.

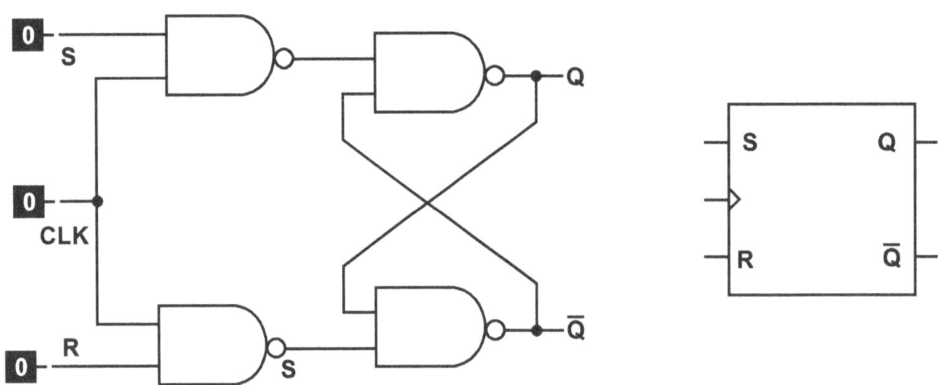

Fig.13.2 : Clocked RS NAND Flip-Flop and symbol

The timed RS latch enables more precise control of the latching process while resolving some of the issues with the conventional RS latch circuit. It is not, however, a comprehensive answer.

The fact that the S and R input levels could readily change while the CLK input is still at a logic 1 level poses a serious issue for this latch circuit. As a result, several states can be changed in the circuit before the CLK input returns to logic 0.

Keeping the CLK mostly at logic 0 and allowing only brief shifts to logic 1 is one technique to reduce this issue. But even with this approach, the latch could still swap states more than once when the clock signal is at logic 1. This signal must be present for a predetermined time for all latches to react to it. Most latches are able to respond to a variety of modifications during this time.

Making sure the latch can only modify its outputs once per a clock cycle is a preferable solution. The circuit that expertly resolves this issue by changing states only on a certain transition, or edge, of the clock signal is illustrated on the following page.

13.3 THE JK FLIP-FLOP

In a timed RS flip flop, a "race" condition results when inputs go to logic 1 at the same time, making it impossible to predict the latch's ultimate state in advance.

We need to find a way to stop one of those inputs from having an impact on the circuit's master latch in order to rule out the possibility of a "race" circumstance. However, if the input logic signals need it, we still want the flip-flop to be able to change states on each falling edge of the CLK input. As a result, the S or R input that needs to be disabled depends on how the slave latch outputs are currently configured.

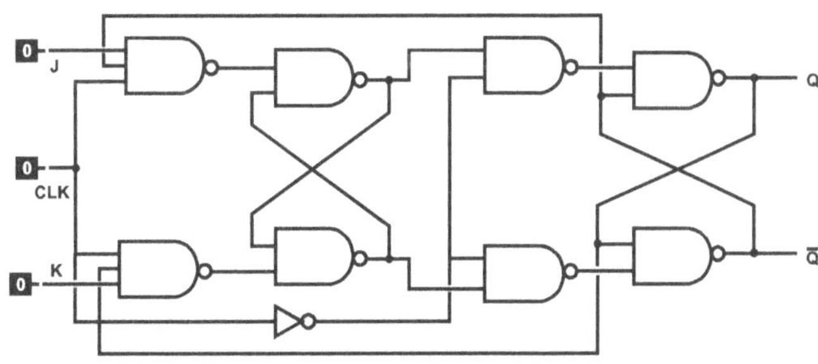

Fig.13.3 : Clocked JK Flip-Flop

The S input cannot make the flip-flop more set than it currently is if the Q output is a logic 1 (indicating that the flip-flop is in the "Set" state). The flip-flop can therefore be disabled in these circumstances without also blocking the S input. In the same way, disabling the R input won't harm the device if the Q output is logic 0 (the flip-flop is set to Reset). If we can do this without too much difficulty, we will have eliminated the "race" condition's issue.

CLK	J	K	Q_{n+1}
0	0	X	Q_n (last state)
↑	0	1	0
↑	1	1	1
↑	1	X	Q_n (last toggle)

Truth Table

111

The resolution is shown in circuit figure 13.3. Two extra connections from the Q and Q outputs back to the original input gates have been added to the RS flip-flop. This is unproblematic because a NAND gate can accept any number of inputs. We rename the flip-flop and the logic inputs to demonstrate that we have achieved this. Instead of S and R, the inputs are now referred to as J and K, respectively. The entire device is referred to as a JK flip-flop.

In this way, the JK flip-flop must be edge triggered. If any level-triggered JK latch circuit's three inputs are held at logic 1, the circuit will oscillate quickly. This is of little use. The T flip-flop must also be edge triggered for the same reason. This is the only technique to guarantee that the flip-flop will switch states just once during any given clock pulse for both types. The JK flip-flop is the flip-flop of choice for the majority of logic circuit designs since its behaviour is entirely predictable in all circumstances. The RS flip-flop is only used in situations where it is certain that R and S cannot both be logic 1.

13.4 MASTER-SLAVE FLIP-FLOP BY JK

Another means of avoiding racing is offered. The slave is first negatively edge-triggered while the master is initially positively edge-triggered. The master therefore reacts to J and K inputs before the slave. The master sets on the leading edge of the positive clock if J = 1 and K = 0. The slave sets when the negative clock edge hits, imitating the master's behaviour because the high Q output of the master drives the J input of the slave.

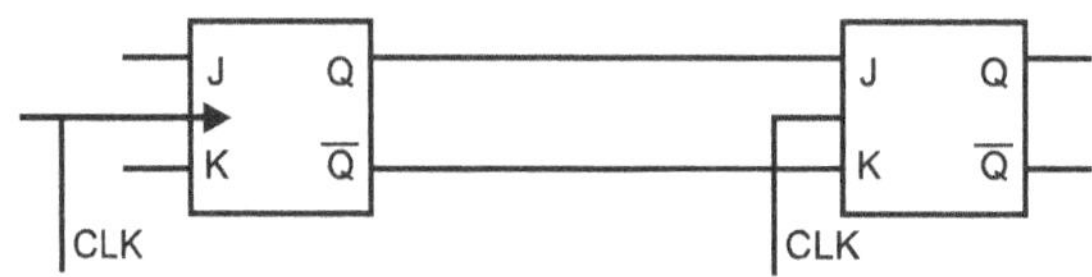

fig. 13.4 : JK Master-Slave flip-flop

The master resets on the clock's leading edge if J = 0 and K = 1. The master's high Q output connects to the slave's K input. The slave must therefore reset when the clock's trailing edge arrives. Once more, the slave has imitated the master.

If both the J and K inputs on the master are high, it toggles on the positive clock edge, and the slave toggles on the opposite clock edge. Therefore, no matter what the master does, the slave replicates it; likewise, if the master sets, the slave also sets; likewise, if the master resets, the slave also resets.

13.5 FLIP-FLOP D-TYPE

There are several situations where an edge-triggered D flip-flop is desirable, although they are not all necessary for safe operation. As shown in fig. 13.2, this can be achieved by employing a D latch circuit as the master section of an RS flip-flop. Both kinds are made commercially available because they are both beneficial.

The logic level present at the D input will always be reflected in the Q output of the D flip-flop when the CLK input is logic 1, regardless of how that level changes. The last state of the D input is trapped and maintained in the latch for use by any other circuits that may require this signal when the CLK input drops to logic 0.

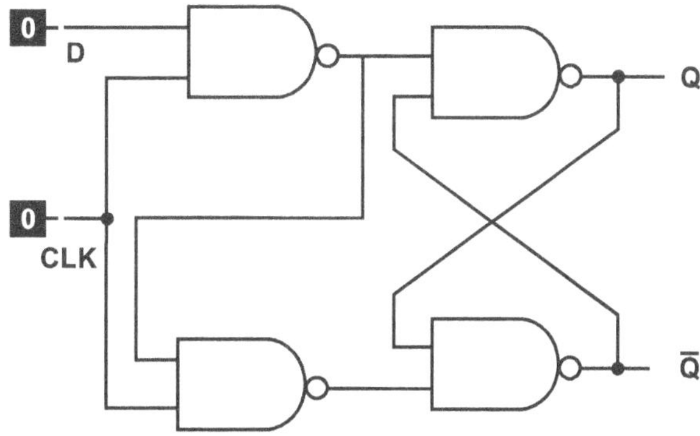
fig. 13.5 : D- Type Flip-Flop

This latch circuit cannot encounter a "race" problem brought on by all inputs being at logic 1, as the sole D input is additionally inverted to supply the signal to reset the latch. The D latch circuit can thus be utilised in any circuit without risk.

At the instant of a rising clock edge and never at any other time, the state of the D input becomes the state of the Q output. Due to the output taking the value of the D input or Data input and delaying it by one clock count, it is known as the D flip-flop. It is possible to think of the D flip-flop as a rudimentary memory cell or delay line.

As the foundation for shift registers, a crucial component of many electrical systems, these flip flops are immensely helpful. The D flip-flop has the benefit over the D-type latch in that it 'captures' the signal when the clock rises high, and further changes to the data line do not affect Q until the next rising clock edge.

13.6 FLIP-FLOPS WITH AN EDGE TRIGGER

One little modification needs to be made to our clock flip-flop. It is important to note that there appears to be some apparent discrepancy in the timing diagram of fig. 13.6 regarding the precise moment at which the D input latches onto Q. What will happen, for instance, if a transition in D occurs at some point during a clock HIGH? The features of the specific electronics being used will determine the response.

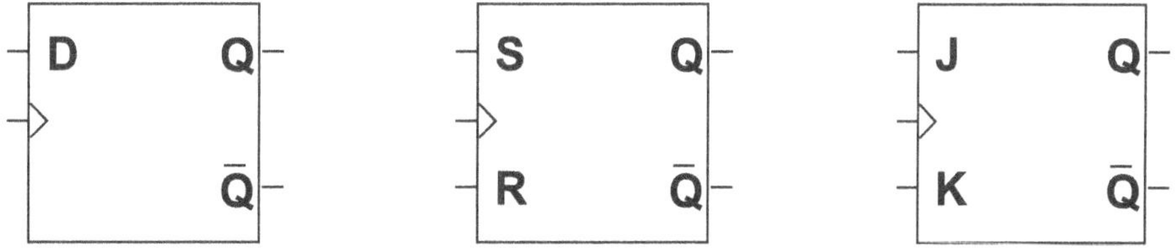
fig. 13.6 : Symbol of different triggered flip flops

The edge-triggered flip-flop is the answer to this problem. Similar to what we have just covered, clock rising or falling edge triggers call for SR-type flip-flops rather than a single

The 7474 IC is a prime example of this in TTL. A negative-edge triggered D-type flip-flop is also frequently used; it latches the D input when the clock changes from HIGH to LOW.

In fig. 13.6, the symbols for these three D-type flip-flops are shown. Be aware that the little triangle at the clock input signifies positive-edge triggering and that the opposite situation is represented by an inverted symbol. The D-type flip-flop has a slightly nicer counterpart called the JK type.

13.7 FLIP-FLOP WITH AN EDGE-TRIGGERED RS

As seen in fig. 13.7, the edge-triggered RS flip-flop is actually made up of two identical RS latch circuits. This is essential for this circuit to work.

The S and R inputs are separated from the input (master) latch if we begin with the CLK input set to logic 0 as seen in the original diagram above. As a result, modifications to the input signals cannot influence how the end outputs behave.

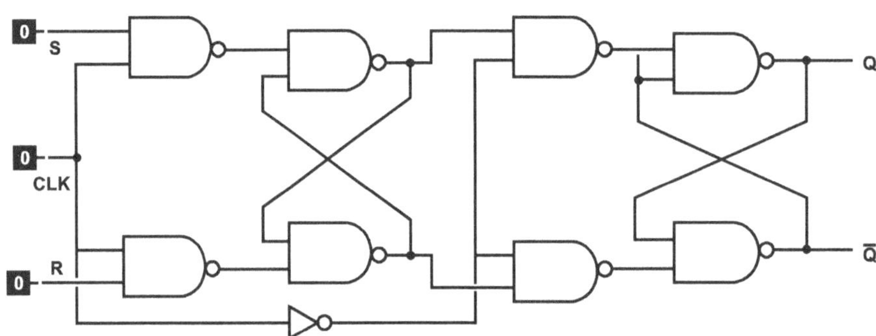

fig. 13.7 : Edge-triggered RS NAND flip-flop

Similar to the single RS latch circuit you previously studied, the S and R inputs can alter the state of the input latch when the CLK signal shifts to logic 1. However, the output (slave) latch's inverted CLK signal prohibits the status of the input latch from having any impact on this situation. As a result, when CLK is at logic 1, the input latch tracks any changes in the R and S input signals, but the Q and $\overline{Q}$ outputs are unaffected.

The S and R inputs are once more isolated from the input latch when CLK once more drops to logic 0. The output latch can now be affected by the input latch's current state thanks to the inverted CLK signal, which also occurs at this moment. The Q and $\overline{Q}$ outputs can only switch states as a result of the CLK signal going from logic 1 to logic 0. Edge-triggered flip-flops are activated by the CLK signal's falling edge, hence the name.

One issue remains: the potential race condition that can arise if both the S and R inputs are at logic 1 when CLK shifts from logic 1 to logic 0. Although we automatically assume in the example above that the race will always end with the master latch in the logic 1 state, this is not a given with real components. As a result, we must find a means to completely avoid race-related situations. So we won't have to work out which gate on the circuit won this particular race.

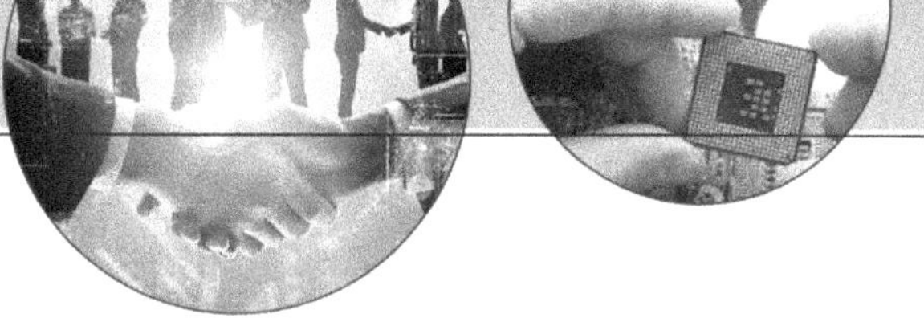

The answer is to give the master latch some additional feedback coming from the slave latch. A JK flip-flop is the name of the resulting circuit.

13.8 APPLICATIONS:

a)   One bit of data, or one binary digit, can be stored in a single flip-flop.

b)   Flip-flops are used to create static RAM, the main type of memory found in registers and many caches used by computers to store numbers.

c)   Any one of the flip-flop varieties can be used to construct any other variety.

d)   The information stored in a number of flip-flops may be anything, including the value of a counter, the state of a sequencer, an ASCII character stored in memory, or any other kind of data.

e)   Electronic logic can be used, for example, to construct finite state machines. The machine's previous state is stored in the flip-flops, and digital logic uses that state to determine the subsequent state.

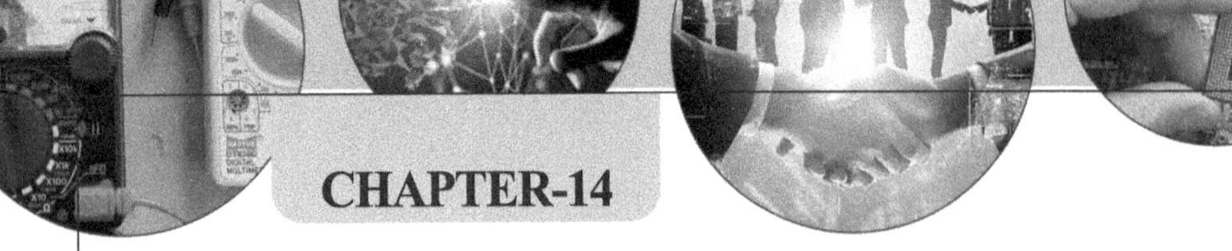

CHAPTER-14

Counters & Registers

Now that we have a basic understanding of flip-flop circuit components, we can use it to create counters and registers with additional functionality.

14.1 COUNTERS

Counters Forward and reverse counting is a typical requirement in digital circuitry. Everywhere you look, you may find digital clocks and watches, timers in a variety of devices like microwaves and VCRs, and counters for different purposes in things like test equipment and cars.

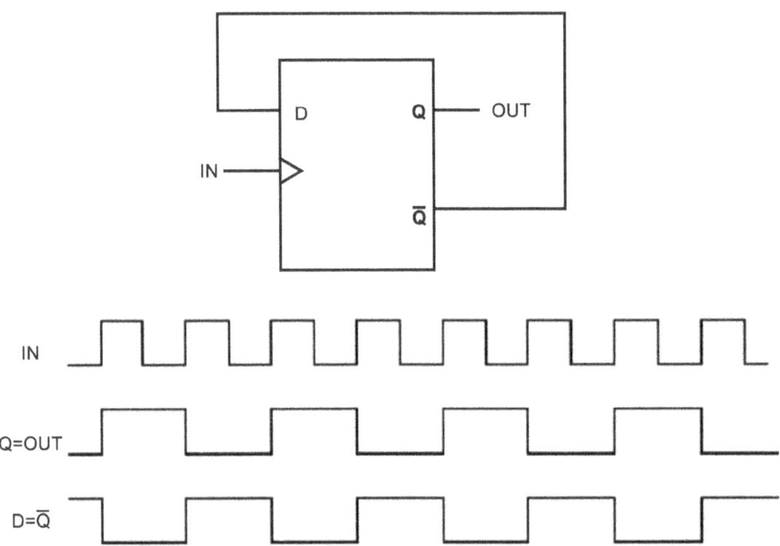

fig. 14.1 : Positive edge-triggered D-type flip-flop connected as divide-by-2 counter.

The edge-triggered D-type flip-flops are very practical and adaptable sequential logic building blocks. The divide-by-2 counter in Fig. 14.1 and the related time diagram represent a straightforward implementation.

14.2 AN ASYNCHRONOUS COUNTER

As seen in fig. 14.2, asynchronous counter flip-flops can be linked in series. The outputs as a result are shown in figure 14.3. if A = 20 and B = 21 and C = 22. The maximum count for this 3-bit counter, then, is 23 - 1 = 7. It is obvious that such a counter might be expanded to contain an infinite number of bits.

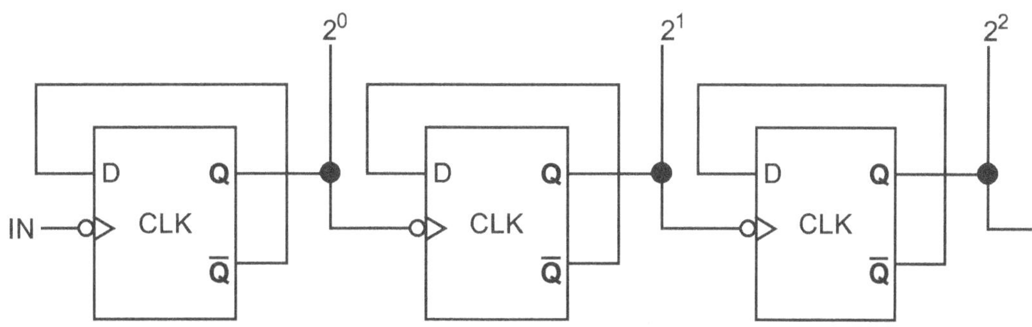

fig. 14.2 Asynchronous counter made from cascaded D-type flip-flops

Asynchronous counters are simple to put together, but they have significant disadvantages in particular situations. Before the right outcome is obtained, the input must specifically pass through the full chain of flip-flops. The two endpoints of the chain, which stand for the LSB and MSB, may eventually handle wholly separate input pulses at high input rates. The synchronous counter, an example of a state machine that we will explore shortly, offers a solution to this problem.

The outputs as a result are shown in figure 14.3. Flip-flop A will switch states each time a negative clock transition occurs. Thus, at point A on the time line, A rises, followed by a drop at point B, a rise at point C, and so on. Take note that the flip-flop A output waveform is half the clock frequency.

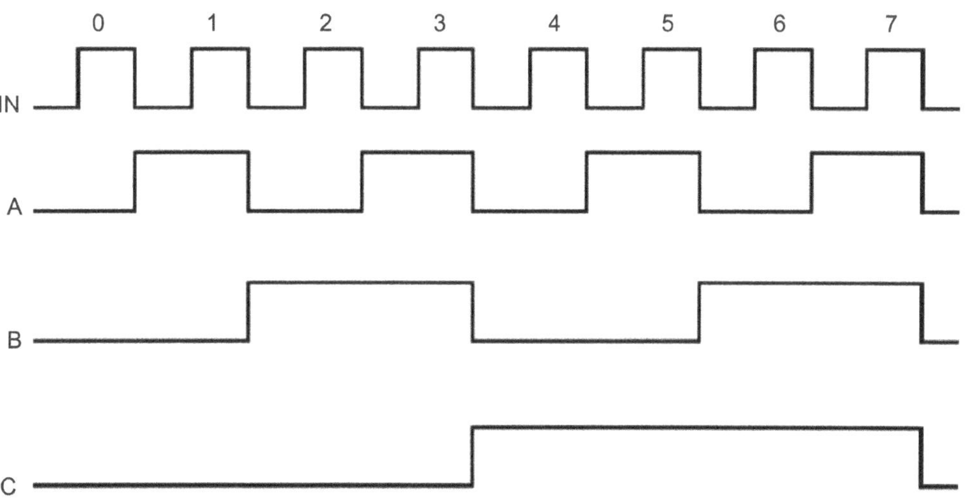

fig.14.3 Waveforms generated by the asynchronous counter.

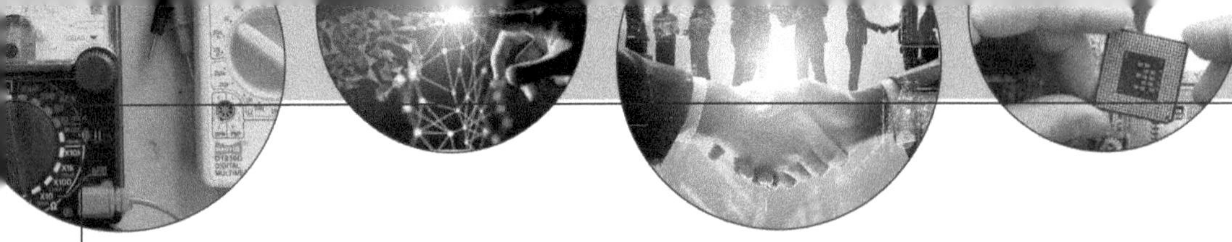

14.2   A SYNCHRONIZED COUNTER

To achieve this, the status of every bit in the output count would change simultaneously. Apply the same clock pulse to all flip-flops in a counter so that they can all work in sync with one another.

We do not, however, want every clock pulse to cause every flip-flop to change states. To govern when each flip-flop may change state and when it cannot, we will thus need to add some controlling gates. For this, we can utilise either RS or JK flip-flops..

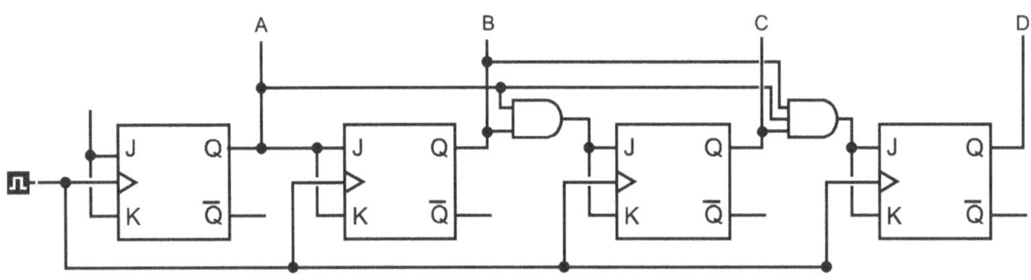

Fig.14.4 Synchronous Counter using JK flip flop

IIn fig. 14.4, we first see that output A must undergo a state change for each pulse of the input clock. just to standardise our flip-flops across the board. But even with JK flip-flops, all we need to do in this case is link this flip-flop's J and K inputs to logic 1, and the activity will be right.

A little more work goes into flip-flop B. Only every other input clock pulse must cause this output to change states. When output A is a logic 1, output B must be prepared to change states, but not when A is a logic 0. This is seen in Fig. 14.5 of the truth table. If we think back to how the JK flip-flop operated, we can see that output B would function appropriately if we connected output A to the J and K inputs of flip-flop B.

Following this logic, only when both A and B are logic 1 may output C change state. We cannot control flip-flop C using only output B; doing so will allow C to change states when the counter is in state 2, allowing it to jump directly from a count of 2 to a count of 7, and then from a count of 10 to a count of 15, which is not a useful method of counting. Therefore, at the inputs of flip-flop C, we will need a two-input AND gate. Because outputs A, B, and C must all be at logic 1 before flip-flip D is permitted to change states, flip-flip D requires a three-input AND gate for control.

States				Count
D	C	B	A	
0	0	0	0	0
0	0	0	1	1
0	0	1	0	2
0	0	1	1	3
0	1	0	0	4
0	1	0	1	5
0	1	1	0	6
0	1	1	1	7
1	0	0	0	8
1	0	0	1	9
1	0	1	0	10
1	0	1	1	11
1	1	0	0	12
1	1	0	1	13
1	1	1	0	14
1	1	1	1	15

Fig.14.5 Synchronous Counter Truth table

14.4 REGISTERS

In all instances when the phrase "register" is used, it refers to a collection of flip-flops that work together to store data as a cohesive unit. This is distinct from a counter, which consists of flip-flops that work together to tabulate data in order to produce new data.

A 4-bit memory is represented in figure 14.6. It can be compared to 4 distinct D-type flip-flops. The crucial feature of a data register of this kind is that each input is simultaneously latched into memory by a single clock cycle.

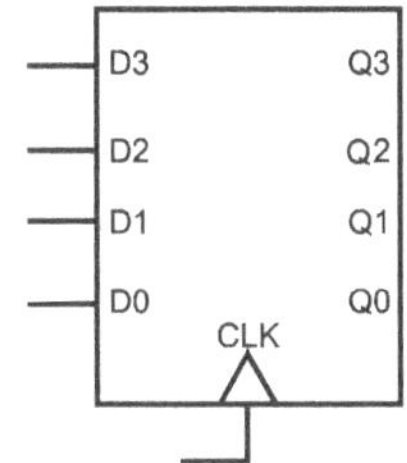

fig. 14.6 4 - bit data register

14.5 SHIFT REGISTER FOR SERIAL TO PARALLEL

A counter can be thought of as a specific form of register that counts occurrences and produces data as a result, as opposed to simply retaining the data or altering how it is handled. However, counters and registers are frequently handled differently. The two are then treated as distinct ideas that share some characteristics and can be used independently or in combination in numerous applications.

An illustration of a 4-bit shift register is shown in figure 14.7. These setups are quite helpful, especially for converting parallel to serial and serial to parallel data. On each clock cycle in the circuit below, a pulse coming in at serial in" would be moved from one flip-flop's output to the next. As a result, after four clock cycles, a serial bit pattern at the input (4 bits long in our example) would show as 4 parallel bits in the outputs Q1–Q4. The serial-to-parallel situation is illustrated by this.

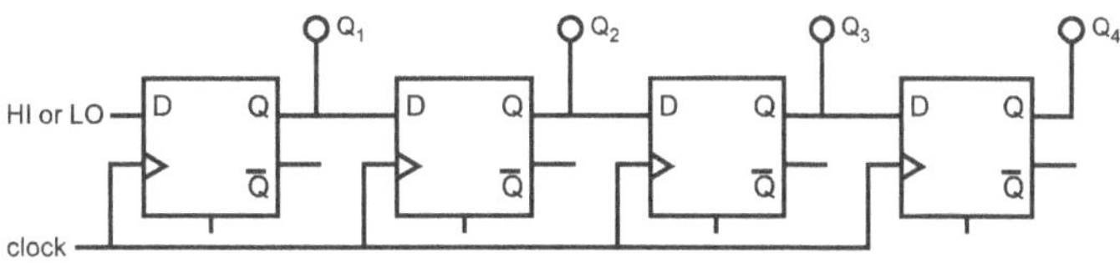

fig. 14.7 4 - bit shift register

14.6 SHIFT REGISTER FOR PARALLEL TO SERIAL

Wherever serial-to-parallel conversion is required, parallel-to-serial conversion is likewise required. But the parallel-in, serial-out register—also known as a parallel-to-serial shift register or shift-out register—is a little more complicated. A modest two-input multiplexer is necessary in front of each input because each flip-flop in the register needs to be able to accept data from either a serial or a parallel source.

The flip-flops are loaded as normal in accordance with a common clock signal, and an additional input line switches between serial and parallel input signals.

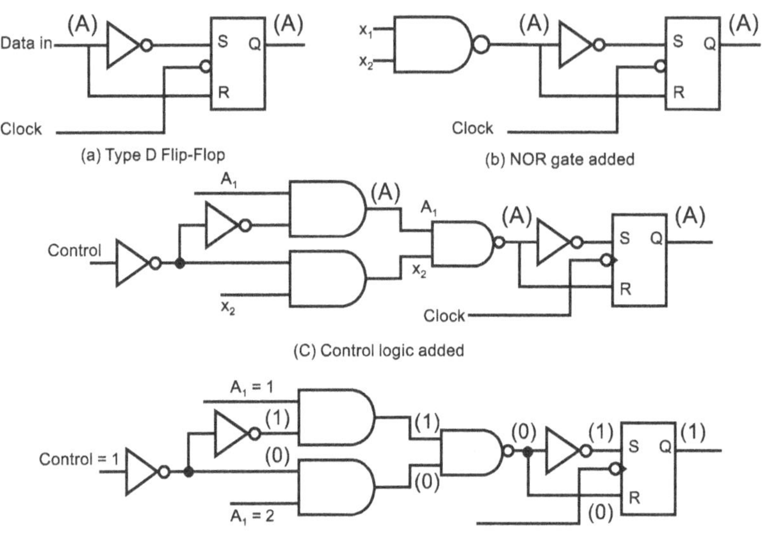

fig. 14.8 Parallel in Serial out

Although there are many alternative methods for entering data in parallel, we will focus on commercially available TTL. In Fig. 14.8, an 8 bit Shift Register 54/74177 is shown. According to the TTL data sheet's functional description, this is an 8-bit shift register with serial and parallel data input and output options. There are eight RS flip-flops visible, each with some logic circuitry attached. Let's examine one of these circuits by beginning with the RS flip-flops and then including logic blocks as necessary to meet our demands.

A D-type flip-flop is created by the clocked RS flip-flop and the associated inverter shown in Fig. 14.8a. The complement of a data bit, such as A, must be present at the input in order for it to be clocked into the flip-flop. For instance, if $A = 0$, then $R = 0$, $S = 1$, and when the clock transitions, a 1 will be clocked into the flip-flop.

Similar to the NOR gate in Fig. 14.8b. A data bit A at the other leg of this NOR gate is simply reversed by the NOR gate if one of its legs is at ground level. For instance, if $A = 1$, then at the NOR gate's output, $A = 0$, enabling the clocking of a 1 into the flip-flop. Data can be entered into this NOR gate from either A1 or A2, two separate sources. while A2 is held at ground, data at A1 can be transferred into the flip-flop; while A1 is held at ground, data at A2 can be shifted into the flip-flop.

Data A1 or data A2 can be chosen by adding two AND gates and two inventors as illustrated in Fig. 14.8C. The top AND gate is activated and the lower AND gate is disabled if the control line is high. As a result, A1 will be visible at the NOR gate's upper leg while it is at ground level for the lower leg. The lower AND gate is enabled when the control line is low, while the higher AND gate is disabled. This enables one to appear at the NOR gate's bottom leg while the gate's higher leg is at ground level.

CHAPTER-15

Multiplexer, Demultiplexer & Encoder, Decoder

15.1 MULTIPLEXER

A multiplexer is a tool that chooses one input from a variety for a single output. Use of an input address is used to make the choice. As a result, a multiplexer can place numerous data bits in a certain order, one at a time, on a single output data line. This is an illustration of serialising parallel data. In order to send N data bits, multiplexing reduces the number of independent lines needed to M Control Lines ($2^m = N$), as opposed to N when serializations is not used.

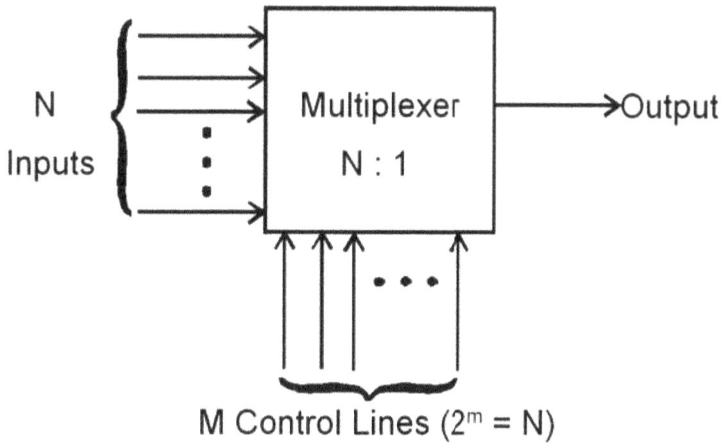

fig. 15.1 Multiplexer

A multiplexer in digital electronics is a grouping of logic gates that produces circuits with two or N inputs (data inputs) and one output. By sending a certain digital word to a different set of inputs (Control Lines), the channel that will be read into the output can be chosen. Multiplexers come in several configurations, such as 2:1, 4:1, 8:1, and 16:1. The 16:1 multiplexer, which has 16 input lines, 4 control lines, and 1 output line, is one of the most well-known multiplexers.

According to Fig. 15.2, IC 74150 is a 16:1 multiplexer based on TTL technology. The input data bits D0 to D15 are connected to pins 1 through 8 and 16 through 23, while the control bits ABCD are connected to pins 11, 13, and 15. The output bit pin is number 10, and the strobe pin is number 9. The multiplexer can be enabled or disabled using the strobe input signal. The multiplexer is enabled via a low strobe. In the logic of a 74150 Multiplexer. Assume we desire a logic circuit similar to the one shown in Figure 15.2. D0 is grounded, followed by D1 and D2 linked to + 5V, and so on. Each time, the data input matches the desired Y output of Table 15.2 exactly.

This method is exactly how the telephone network combines several audio signals onto a single pair of lines, making it simple to segregate many phone conversations so that each person's voice only reaches the designated receiver.

Most people have heard about T1 telephone lines thanks to the development of the Internet and the World Wide Web. This multiplexing technique allows up to 24 separate telephone conversations to be transmitted on a T1 line.using exactly this technique, and is readily able to separate many telephone

Dynamic memory uses the same address lines for both row and column addressing in computers, which is a highly typical application for this kind of circuit. The memory's row address is first chosen using a set of multiplexers, and then the column address is chosen. This design makes it possible to fit more memory into the computer while using fewer copper traces to connect the memory to the rest of the computer's electronics. This circuit is generally known as a data selector in such a scenario.

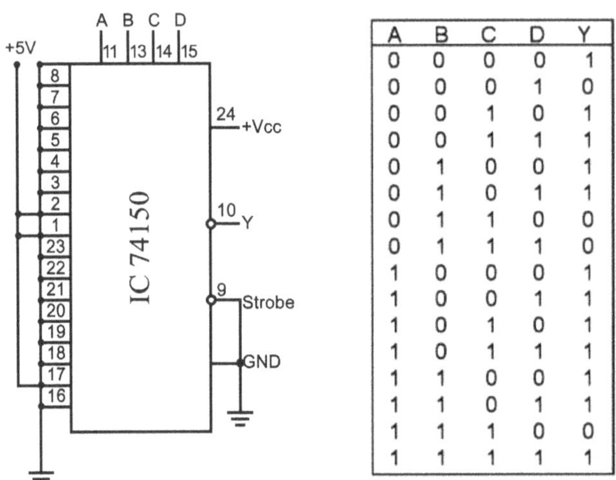

A	B	C	D	Y
0	0	0	0	1
0	0	0	1	0
0	0	1	0	1
0	0	1	1	1
0	1	0	0	1
0	1	0	1	1
0	1	1	0	0
0	1	1	1	0
1	0	0	0	1
1	0	0	1	1
1	0	1	0	1
1	0	1	1	1
1	1	0	0	1
1	1	0	1	1
1	1	1	0	0
1	1	1	1	1

fig. 15.2 74150 Multiplexer logic and Truth Table

15.2: DEMULTIPLEXER

The opposite operation is carried out by a demultiplexer, which sends a single input to one of numerous potential outputs. Once more, an address is used to choose the output line.

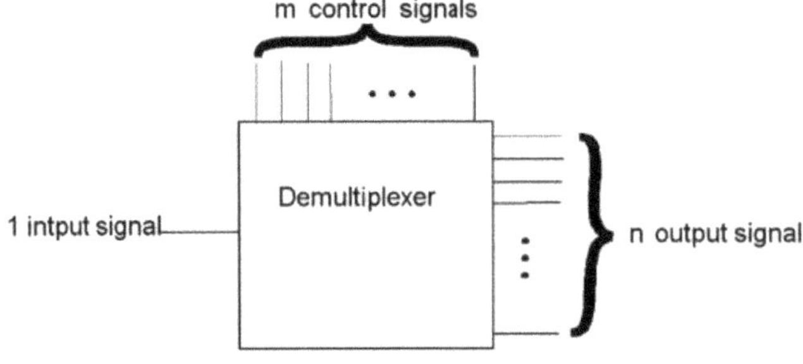

fig. 15.3 Demultiplexer

The pinout of the 74154's 1-to-16 demultiplexer is shown in fig. 15.4. Pins 1 through 11 and 13 to 17 are for the output bits yo to y 15', while pin 18 is for the input data D. ABCD control bits are Pins 20 to 23. Strobe pin 19 is again an active-Iow input. The pins 24 and 12 are used for ground and V cc, respectively.

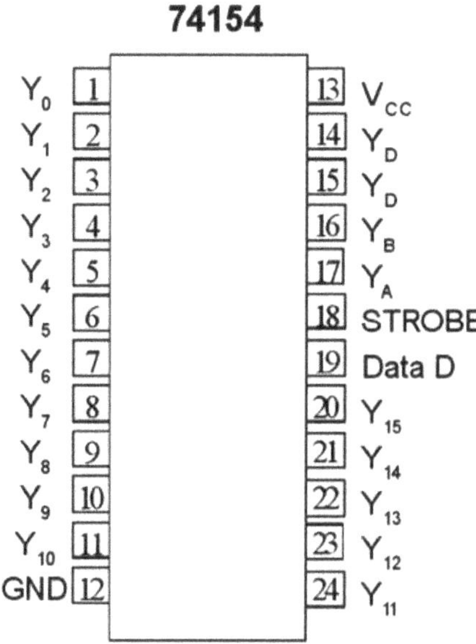

74154

fig. 15.4 Pinout Diagram of 74154

The data input line serves as a circuit enabler; regardless of the binary input number, no output will display activity if the circuit is disabled. The quantity of transmission lines needed can be decreased by using a Multiplexer-Demultiplexer pair to convert data into serial form for transmission. The Multiplexer and Demultiplexer at either end share the address bits.

15.3 ENCODER

The term "multiplexing" is also occasionally used to describe the encoding process in digital circuits, which is essentially the creation of a digital code to identify which of several input lines is active. An encoder or multiplexer is a digital integrated circuit (IC) that generates a digital code depending on which of its several digital inputs is active.

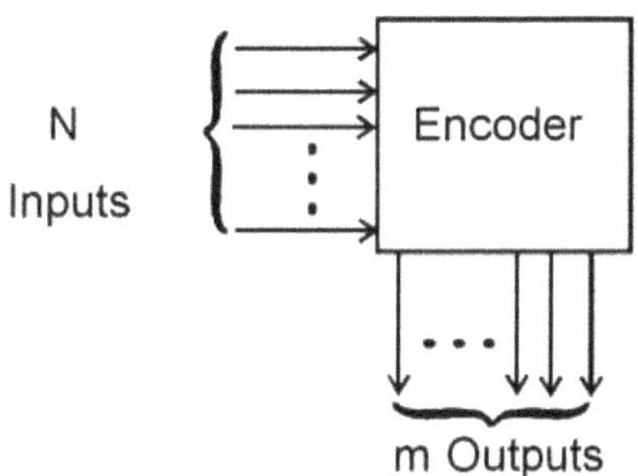

15.4 BCD TO DECIMAL ENCODER

The four-digit binary equivalents of the 10 decimal digits (0, 1,..., 9) are used to represent them. The Decimal-to-BCD encoder should therefore have 10 inputs and 4 outputs. Figure 15.5, which shows the logic symbol for this 10-to-4-line decoder, and Table 15.6, which lists the corresponding conversion table, both show A3 as the most significant bit.

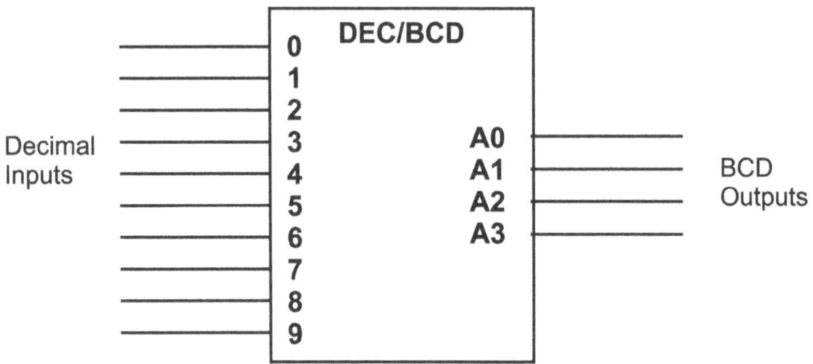

fig. 15.5 Logic symbol for Decimal-to-BCD Encoder

Decimal Digit	BCD Code			
	A_3	A_2	A_1	A_0
0	0	0	0	0
1	0	0	0	1
2	0	0	1	0
3	0	0	1	1
4	0	1	0	0
5	0	1	0	1
6	0	1	1	0
7	0	1	1	1
8	1	0	0	0
9	1	0	0	1

fig. 15.6 Decimal-to-BCD code table

The definition of the MSB by the formula $A3 = 8 + 9$ is shown in Fig. 15.6.
Similarly, the functions may be define $A2$, $A1$, and $A0$.

As shown in fig. 15.7, we can now put into practise the logic circuit for the decimal-to-BCD decoder.
It should be noted that there is no need for a connection for the decimal digit zero in this instance because there are no HIGH inputs, hence all of the BCD outputs are LOW.

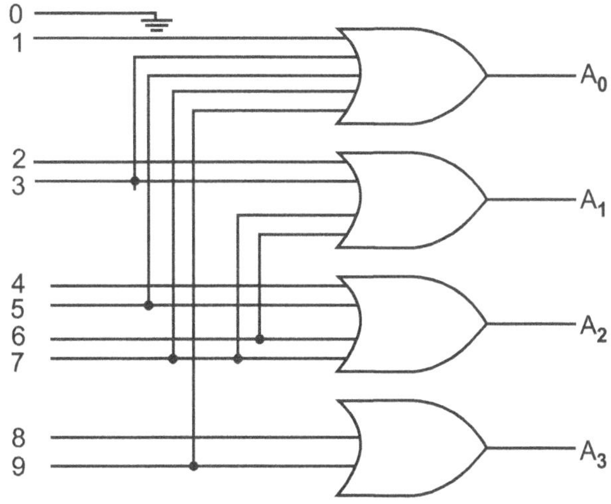

fig. 15.7 Logic circuit for Decimal-to-BCD encoder

15.5 DECODER

According to the digital code contained at the binary-weighted inputs of the decoding circuit or decoder, decoding is the process of activating one of many mutually exclusive output lines, is also known as "demultiplexing" in the context of digital electronics. Decoders or demultiplexers are essentially digital integrated circuits (ICs) that accept digital codes made up of two or more bits as inputs and, based on the value of the code, enable or activate one of their several digital output lines.

With one exception, a decoder is comparable to a demultiplexer except that there is no data input. The control bits (m control lines) are the only inputs.

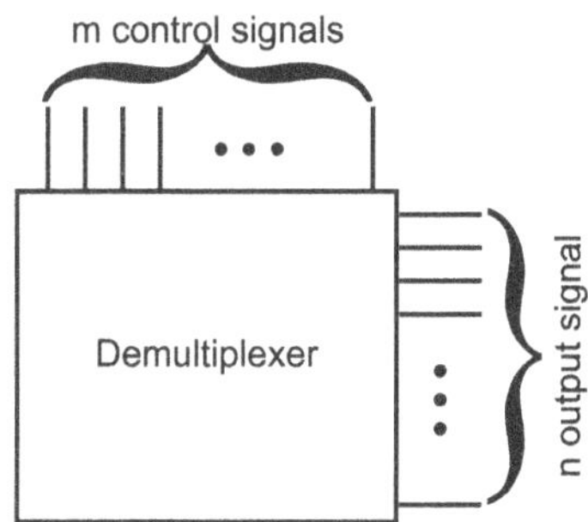

fig. 15.8 Demultiplexer, block diagram

15.6 THE DECODER FOR BCD TO DECIMAL

Each BCD code is translated into its decimal counterpart by the BCD-to-decimal decoder. The method utilised is quite comparable to the method used to create the 3-line-to-8-line decoder. The decoding functions for the BCD-to-decimal decoder are listed in Fig. 15.9, again assuming that active high outputs are necessary.

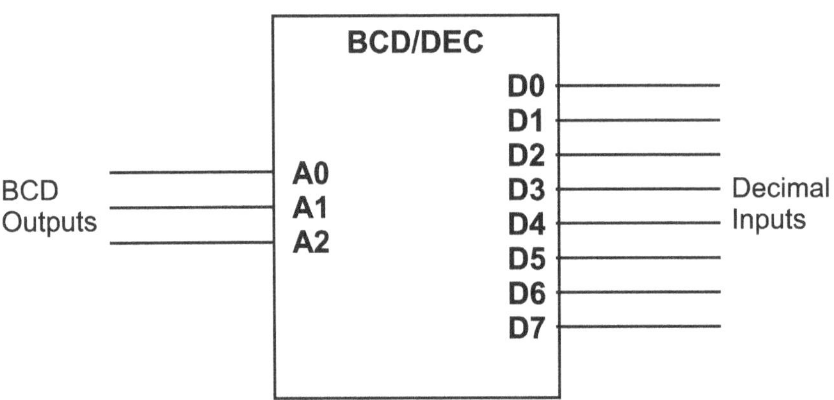

fig. 15.9 Logic Symbol for 3-line-to-8-line decode

Decimal Digit	Binary Inputs			Logic Function	outputs							
					D_0	D_1	D_2	D_3	D_4	D_5	D_6	D_7
0	0	0	0	$\bar{A}_2\bar{A}_1\bar{A}_0$	1	0	0	0	0	0	0	0
1	0	0	1	$\bar{A}_2\bar{A}_1A_0$	0	1	0	0	0	0	0	0
2	0	1	0	$\bar{A}_2A_1\bar{A}_0$	0	0	1	0	0	0	0	0
3	0	1	1	$\bar{A}_2A_1A_0$	0	0	0	1	0	0	0	0
4	1	0	0	$A_2\bar{A}_1\bar{A}_0$	0	0	0	0	1	0	0	0
5	1	0	1	$A_2\bar{A}_1A_0$	0	0	0	0	0	1	0	0
6	1	1	0	$A_2A_1\bar{A}_0$	0	0	0	0	0	0	1	0
7	1	1	1	$A_2A_1A_0$	0	0	0	0	0	0	0	1

Fig.15.10 Decoding functions and truth tables for the 3-line-to-8-line decode

Eight (2^3=8) decoding logic gates are needed in order to decode all three-bit combinations. The 3-line-to-8-line decoder is the name given to this type of decoder since it has 3 inputs and 8 outputs. Assuming we need ACTIVE HIGH outputs, let's think about the architecture of such a decoder. In other words, the decoder outputs "1" for a specific set of inputs. Fig. 15.10, which lists the decoding operations and truth tables for the 3-line-to-8-line decoder, serves as an example.

We can implement the SOP logic circuit and create a decoder based on each logic function. This is seen in Fig. 15.11 below.

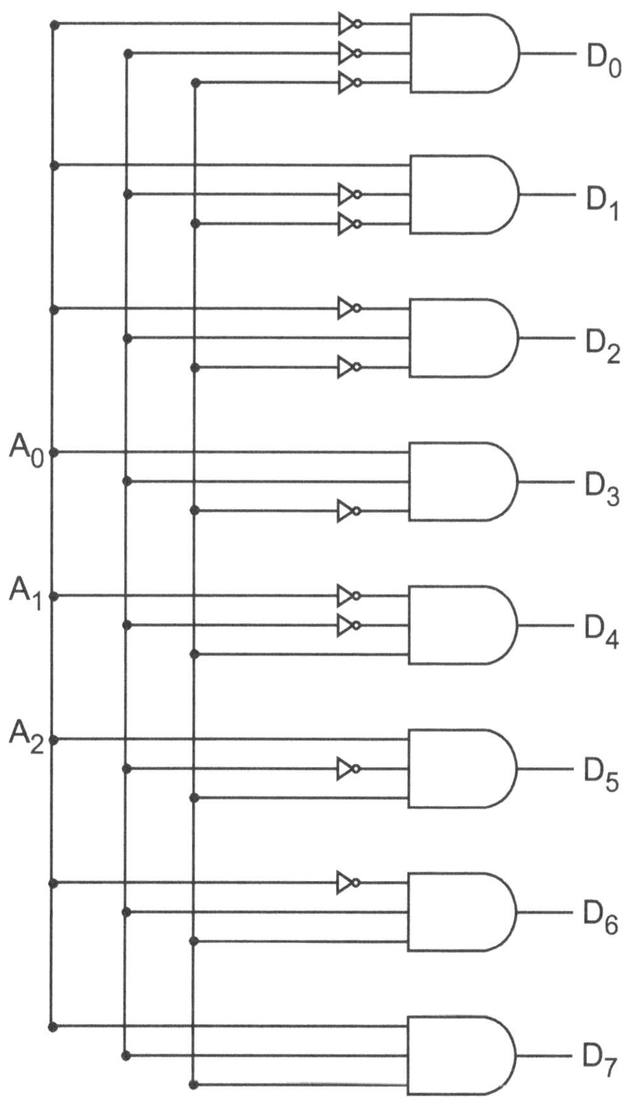

fig. 15.11 Circuitry for 3-lines-to-8-lines decoder

15.7   7-SEGMENT BCD TO DECODER

The 74LS47 is a low power Schottky BCD to 7-Segment Decoder Driver with seven AND-OR-INVERT gates, NAND gates, and input buffers. They provide active LOW, HIGH sink current outputs that can be used to directly drive indicators. To make BCD data and its complement available to the seven decoding AND-OR-INVERT gates, seven NAND gates and one driver are wired in pairs.

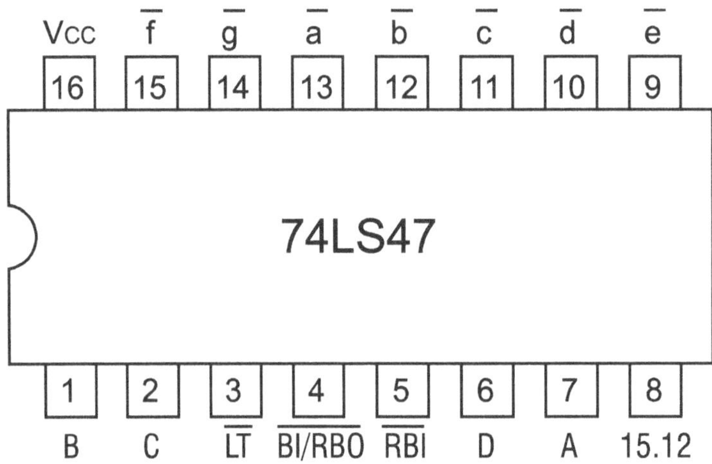

Lamp test (LT) blanking input(BI) / ripple-blanking output (RBO) and ripple-blanking input are provided by the final NAND gate, along with three input buffers. Depending on the status of the auxiliary inputs, the circuits decode the 4-bit binary-coded-decimal (BCD) data and use it to operate a 7-segment display indicator. Figure 15.12, the truth table, displays the relative positive-logic output levels as well as the prerequisites at the auxiliary inputs.

The 74LS47's output configurations are made to handle the comparatively high voltages needed for 7-segment indicators. These outputs have a maximum reverse current of 250 mA and can withstand 15 V. The 74LS47 high performance output transistors can be used to drive indicator segments that need up to 24 mA of current. Display patterns for BCD input counts greater than nine are distinctive symbols to verify input circumstances.

Automatic leading and/or trailing edge zero-blanking control (RBI and RBO) is a feature of the 74LS47. Any time the BI/RBO node is at a HIGH level, the lamp test (LT) may be run This device also includes an override blanking input (BI), which can be used to stop the outputs or regulate the lamp intensity by varying the frequency and duty cycle of the BI input signal.

Ability to Modulate the Intensity of a Lamp (BI/RBO)

Available Collector Outputs Limit high-speed termination effects with leading/trailing zero suppression input clamp diodes during lamp testing.

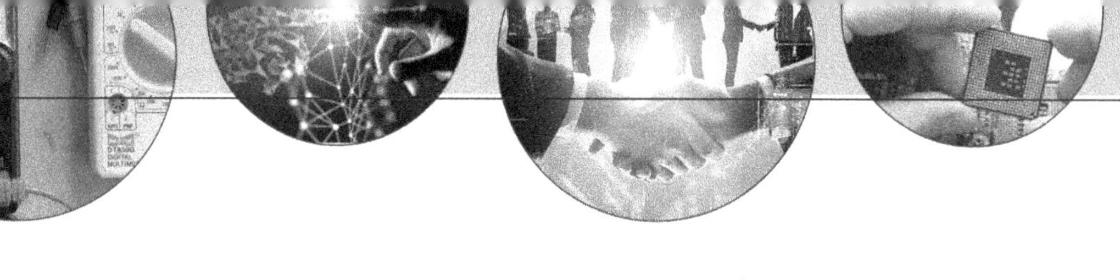

Decimal or Function	$\overline{LT}$	RBI	D	C	B	A	$\overline{BI/RBO}$	$\overline{a}$	$\overline{b}$	$\overline{c}$	$\overline{d}$	$\overline{e}$	$\overline{f}$	$\overline{g}$	NOTE
0	H	H	L	L	L	L	H	L	L	L	L	L	L	H	A
1	H	X	L	L	L	H	H	H	L	L	H	H	H	H	A
2	H	X	L	L	H	L	H	L	L	H	L	L	H	L	
3	H	X	L	L	H	H	H	L	L	L	L	H	H	L	
4	H	X	L	H	L	L	H	H	L	L	H	H	L	L	
5	H	X	L	H	L	H	H	L	H	L	L	H	L	L	
6	H	X	L	H	H	L	H	H	H	L	L	L	L	L	
7	H	X	L	H	H	H	H	L	L	L	H	H	H	H	
8	H	X	H	L	L	L	H	L	L	L	L	L	L	L	
9	H	X	H	L	L	H	H	L	L	L	H	H	L	L	
10	H	X	H	L	H	L	H	H	H	H	L	L	H	L	
11	H	X	H	L	H	H	H	H	H	L	L	H	H	L	
12	H	X	H	H	L	L	H	H	L	H	H	H	L	L	
13	H	X	H	H	L	H	H	L	H	H	L	H	L	L	
14	H	X	H	H	H	L	H	H	H	H	L	L	L	L	
15	H	X	H	H	H	H	H	H	H	H	H	H	H	H	
B1	X	X	X	X	X	X	L	H	H	H	H	H	H	H	B
REI	H	L	L	L	L	L	L	H	H	H	H	H	H	C	C
LT	L	X	X	X	X	X	H	L	L	L	L	L	L	L	D

H= HIGH Voltage Level
L= LOW Voltage Level
X=Immaterial

fig. 15.12 4-bit BCD 74LS47 Truth table

15.8   LED DISPLAY WITH 7 SEGMENTS

7-segment LED displays come in two main varieties. The cathodes of all the LEDs are connected in a common cathode display, and HIGH voltages are used to illuminate each segment individually.

A common anode display cannot be driven by the 4511 because it is meant to drive a common cathode display.

The IC-4511 employs CMOS technology. For 16-pin DIL (dual in line) devices, a power supply with a voltage range of 3 to 15 volts is necessary (see fig. 15.13).

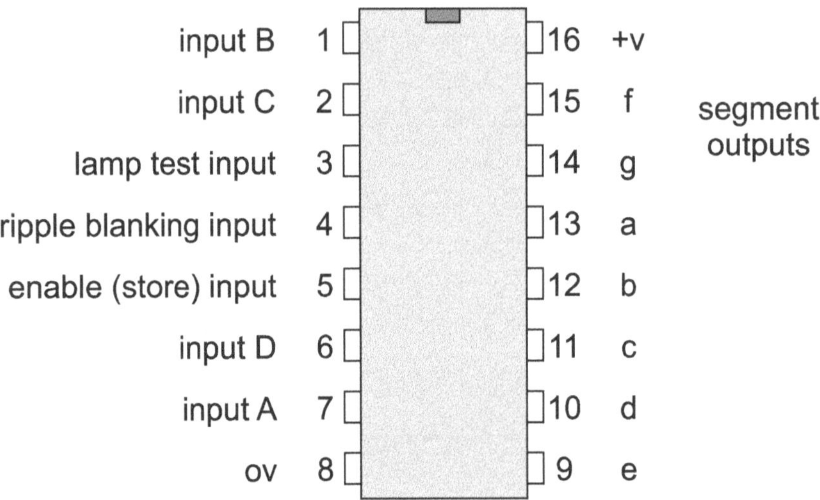

fig. 15.13 IC 4511 Pin connections

A BCD to 7-segment decoder's job is to transform the output logic states of a BCD counter, like the 4510, into a format that can drive a 7-segment display. The display is simple to understand and shows the decimal numerals 0 through 9.

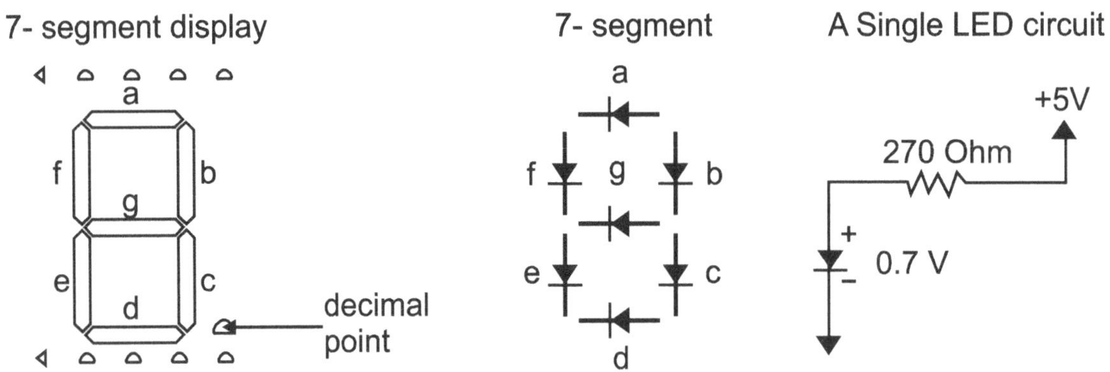

fig. 15.14, 7-Segment display components

A 7-segment display's individual segments are designated by the following letters: Combination of 7-PN junction diodes (7- LED) is shown in Fig. 15.13.

The outputs of the 4511, when configured properly, correspond to the truth table shown in figure 15.14.

BCD INPUTS				SEGMENT OUTPUTS							DISPLAY
D	C	B	A	a	b	c	d	e	f	g	
0	0	0	0	1	1	1	1	1	1	0	0
0	0	0	1	0	1	1	0	0	0	0	1
0	0	1	0	1	1	0	1	1	0	1	2
0	0	1	1	1	1	1	1	0	0	1	3
0	1	0	0	0	1	1	0	0	1	1	4
0	1	0	1	1	0	1	1	0	1	1	5
0	1	1	0	0	0	1	1	1	1	1	6
0	1	1	1	1	1	1	0	0	0	0	7
1	0	0	0	1	1	1	1	1	1	1	8
1	0	0	1	1	1	1	0	0	1	1	9

fig. 15.14. Truth Table

The existence of tails on the 6 and 9 is up for debate. The display from the 4511 is without tails. The outputs of the 4511 are all zeros, and the display is blank if any other binary values other than 1 0 0 1 are connected to the inputs of the device.

CHAPTER-16

Logic Minimization & Karnaugh Maps

A good logic expression can always be created by making an OR out of all the ANDs of the input variables for which the output is true (Q = 1). However, such a process could result in a very long and complicated expression for a truth table of any kind, which might be unnecessarily complex to implement with gates.

Why learn about Karnaugh maps? Similar to Boolean algebra, the Karnaugh map is a digital logic simplification tool. In most instances, the Karnaugh Map will speed up and make logic simpler. For a task requiring two or fewer Boolean variables, Boolean simplification is actually quicker than the Karnaugh map. At three variables, it is still extremely useful but a little slower. Boolean algebra becomes tiresome around four input variables. Karnaugh maps are quicker and simpler. Karnaugh maps can be used with up to eight variables and perform effectively with up to six input variables.

16.1 HOW TO MAKE TRUTH TABLE?

Here we will study about 2, 3 and 4 variables truth tables.

Two Variable Truth Table

Let us consider two variables A & B are as:
$$2^2 = 4$$
So, there will be two 0's and two 1's in the first column(A) of Truth table shown in fig.16.1

A	B	Y
0	0	0
0	1	0
1	0	1
1	1	1

fig. 16.1 Two-Variable truth table

In second column(B), there will be single 0's and single 1's and out put Y will be according to consideration.

Three Variable Truth Table

Let us consider two variables A , B and C are as:

$$2^3 = 8$$

So, there will be four 0's and four 1's in the first column(A) of Truth table shown in fig.16.2

A	B	C	y
0	0	0	0
0	0	1	0
0	1	0	1
0	1	1	0
1	0	0	0
1	0	1	0
1	1	0	1
1	1	1	1

fig. 16.2 Three-Variable truth table

In second column(B), there will be two 0's and two 1's; In third column(C), there will be single 0's and single 1's. Y will be according to consideration.

Four Variable Truth Table

Let us consider two variables A, B, C and D are as:

$$2^4 = 16$$

So, there will be eight 0's and eight 1's in the first column(A) of Truth table shown in fig.16.3
For second column(B), there will be four 0's and four 1's.
For third column(C), there will be two 0's and two 1's.
For fourth column(D), there will be single 0's and single 1's.
Y (outputs)will be according to consideration.

A	B	C	D	y
0	0	0	0	0
0	0	0	1	1
0	0	1	0	0
0	0	1	1	0
0	1	0	0	0
0	1	0	1	0
0	1	1	0	1
0	1	1	1	1
1	0	0	0	0
1	0	0	1	0
1	0	1	0	0
1	0	1	1	0
1	1	0	0	0
1	1	0	1	0
1	1	1	0	1
1	1	1	1	0

fig. 16.3 Four-Variable truth table

16.2 HOW TO MAKE K-MAP?

Here we will draw 2, 3 and 4 variables K-Map as given below in fig.16.4.

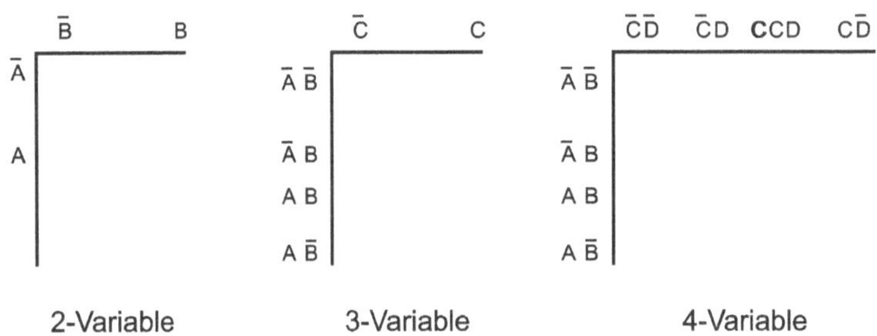

fig. : 16.4   2,3,4 Variable K-Map

16.3 HOW TO MAKE K-MAP WITH THE HELP OF TRUTH TABLE?

Let's start with a simple example. The table below gives an arbitrary truth table involving 2 logic inputs as shown in fig.16.5a .

A	B	B
0	1	0
0	0	0
1	0	1
1	1	1

(a)

(b)

fig.: 16.5, Contruction of K-Map

There are two overall strategies:

1. Copy an expression from the truth table directly. Use Boolean algebra to simplify, if necessary.
2. Employ K-map, or Karnaugh mapping. This only applies if there are no more than four inputs.

The truth table in fig. 16.5a has two 1s in it. Both of them must be present on the K-map. In the third row of the truth table above, find the first 1.

- Take note of the Truth Table Address in AB

- Find the K-map cell with the same address.

- Enter 1 into that cell.

Repeat the procedure for the 1 in the truth table's final line.

- Locate the cell with the same address on the K-map after taking a note of the truth table AB address. place a 1 in that cell.

16.4 HOW TO SOLVE K-MAP ?

Following are the guidelines for minimising Karnaugh Maps:

I)   All of the 1's in the Karnaugh Map are called min terms.

ii) The 1's can be reduced in groups of 2, 4 & 8.

iii) Min terms that are next to each other horizontally or vertically, can be grouped together.

iv) Min terms that have been grouped in a Karnaugh Map, can be reduced to the boolean terms that are common with all the terms in the group.

v)   Min terms that cannot be grouped together, cannot be reduced.

vi) You may use a min term for grouping more than once.

vii) All 1's must be accounted for

Example:

Write the Boolean equation for the outcome for the Truth table below after moving the outputs to the Karnaugh. Let Y = output, $\bar{A}$ = 1; A = 0; and B = 1; $\bar{B}$ = 0.

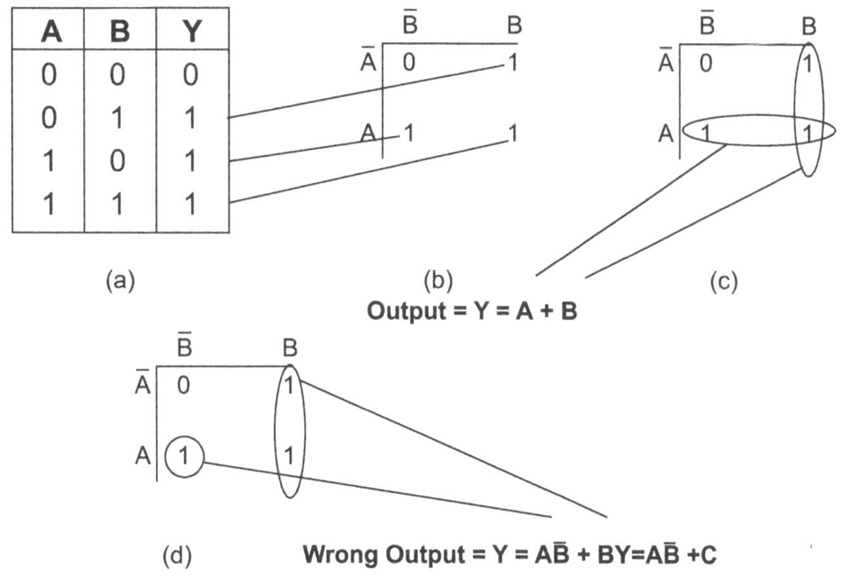

fig. 16.6   Solution of K-Map

Solution:

Transfer the 1s from the Truth table's positions to the K-map's matching spots.

the two 1s in the column beneath "B=1" together (in a circle).

Right after A=1, group (circle) the two 1s in the row.

Put B in the product term for the first group.

Enter "A" as the second group's product word.

List the sums of the aforementioned two terms in your writing. Output: A + B

The K-map's simplest or least expensive solution is shown in Figure 16.6 c. The alternative in fig.16.6 d is less preferable. After combining the two 1s, we make the undesirable mistake of creating a group of 1-cell.

Simplify the logic diagram

- Describe the original logic diagram's Boolean expression.
- The Karnaugh map should be updated with the product terms.
- Form groups of cells as in fig. 16.6 c.
- Write Boolean expression for groups as in fig. 16.6 c
- Draw simplified logic diagram

1) The truth table in fig. 16.6 (a) will be used to create the following logic diagram:

Output $= Y = \overline{A}B + A\overline{B} + AB$

2) The logic diagram will look like this: Output $= Y = A + B$ in accordance with the K-Map illustrated in fig. 16.6 (c).

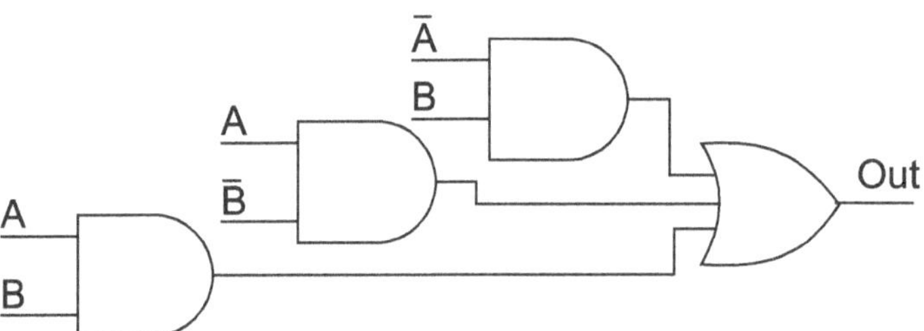

The following ideas, as seen below, will assist you in minimizing any K-Map:

1) Pair

It is a collection of nearby two 1s. In each of the four variables, the two product terms make a pair of two terms. K-Maps, which are simplified to $\overline{A}\,B\overline{C}$ and $A\overline{C}\,\overline{D}$. in Figure 16.7(a,b), are illustrated.

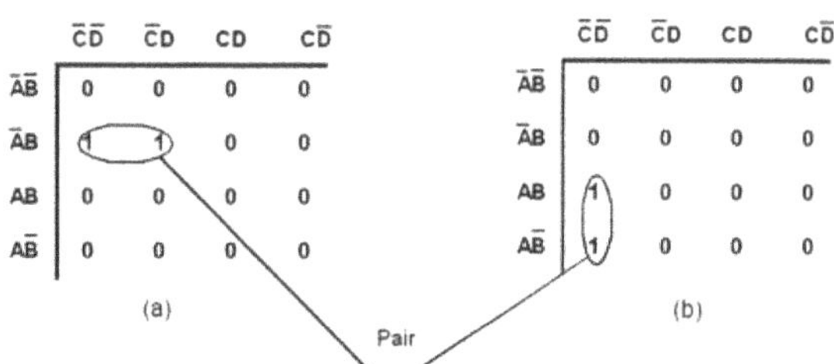

fig. 16.7 pair in 4-variable K-Map

The boolean equation
$$Y = \overline{A}B\,(\overline{C}\,\overline{D} + \overline{C}D)\ (\text{fig.}16.7\,a)$$
$$= \overline{A}B\overline{C}(\overline{D}+D)$$
$$= \overline{A}B\overline{C}(0+1)$$
$$= \overline{A}B\overline{C}(1)$$
$$= \overline{A}B\overline{C}$$

Repeat as follows for fig.16.7 b.
The boolean formula $Y = A\overline{C}\,\overline{D}$

2) Quad

It has four adjacent 1s in a row and two variables. Two groupings that cover each 1 are shown in Fig. 16.8. These classifications will all be reduced to two terms. and is reduced to $A\overline{C}$ and $C\overline{D}$.

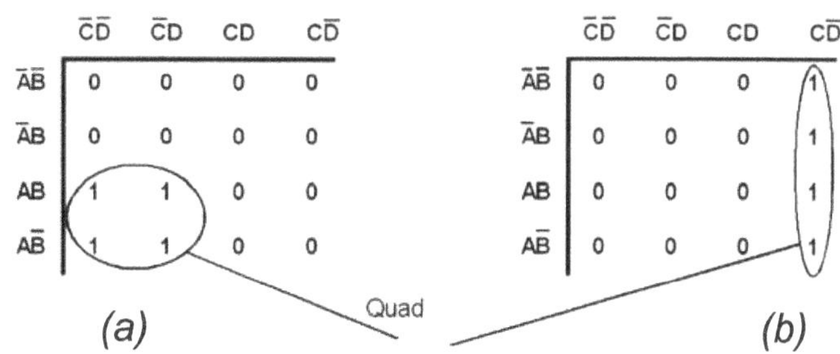

fig. 16.8 (a) (b) Quad in 4-variable K-Map

The boolean equation
$$Y = (AB + A\overline{B})(\overline{CD} + \overline{C}D)\ (\text{fig.}16.8\,a)$$
$$= A(B+\overline{B})\overline{C}(\overline{D}+D)$$
$$= \overline{A}(1+0)\overline{C}(0+1)$$
$$= A\overline{C}$$

Repeat as follows for fig. 16.8b
Boolean expression $Y = C\overline{D}$

3) Octate in K-Map

There are three variables and eight adjacent 1s in the group. Two groupings that cover each 1 are shown in Fig. 16.9. Each of these groupings will be implified to A and C and reduced to three terms.

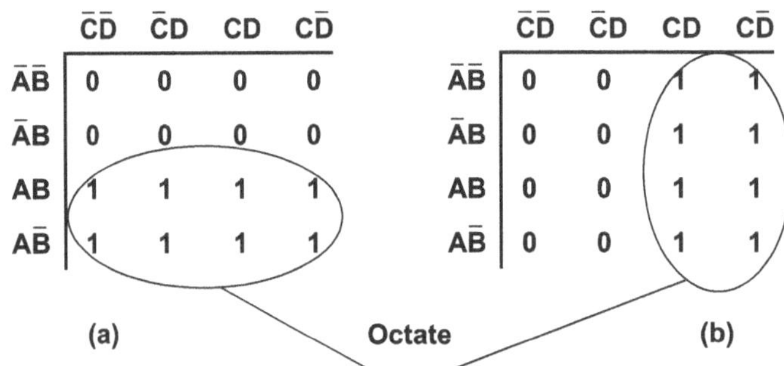

fig. 16.9 Octate in 4-variable K-Map

Boolean expression
$$Y = (AB + A\overline{B})(\overline{C}\,\overline{D} + CD + CD + C\overline{D}) \qquad (fig.16.9a)$$
$$= A(B + \overline{B}) \qquad (1)$$
$$= A(1 + 0)$$
$$= A$$

Repeat as follows for Fig. 16.9b.
Boolean formula $Y = C$

3) Overlapping in K-Map

If at all feasible, overlap your groups. In other words, repeat the number 1 to create the biggest group possible. Map the octate and quad in fig. 16.10a, then select the lower two cells of the left group of eight as a sharing the two with two additional people from the other group to create a group of four in fig. 16.10a. A simpler outcome results from covering these two with a group of eight and four. Due to the existence of two groups, $\overline{C}+AD$ will have two Sum-of-Products results.

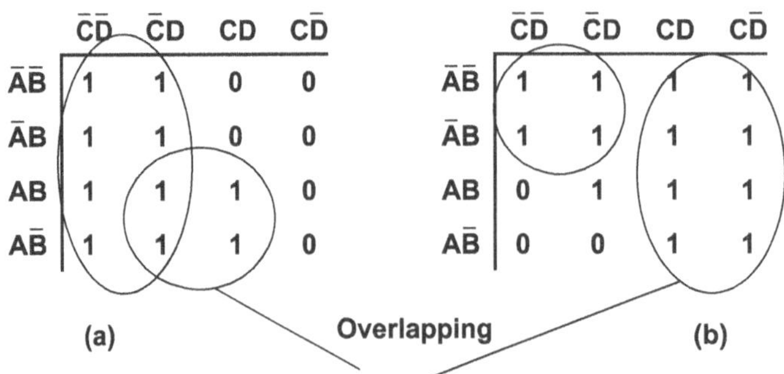

fig. 16.10 Overlapping in 4-variable K-Map

The Boolean Expression

$$Y = (\overline{C}\,\overline{D}+\overline{C}D) + (AB+A\overline{B})(\overline{C}D+CD)$$
$$= \overline{C}(\overline{D}+D) + A(B+\overline{B})D(\overline{C}+C)$$
$$= \overline{C}(0+1)+A(1+0)D(0+1)$$
$$= \overline{C}+AD$$

As shown below, repeat for fig 16.10 (b).

The Boolean Expression Y $= \overline{A}\,\overline{C}+C$

4) Rolling of K-Map

Think of the quartet as being formed by the adjacent cells after rolling the map's ends into a cylinder. The eight and four pieces are typically marked as fig. 16.11, left-right, top-bottom.

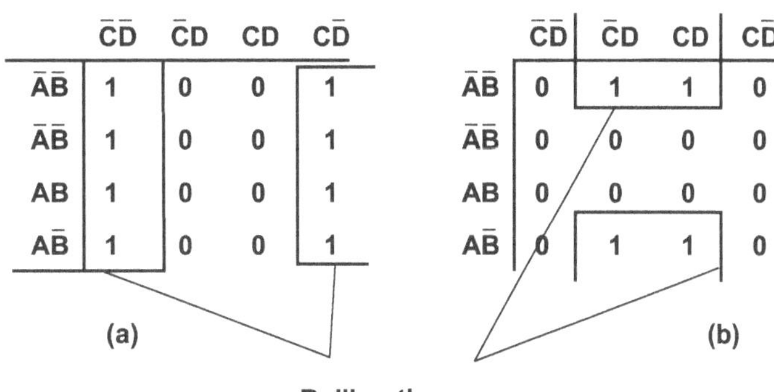

Rolling the map

fig. 16.11 Overlapping in 4-variable K-Map

The Boolean Expression
(fig.16.11a)

$$\mathbf{Y = (\overline{A}\,\overline{B} + \overline{A}B+AB + A\overline{B})(\overline{C}\,\overline{D}+ C\overline{D})}$$
$$= (1)\,\overline{D}\,(\overline{C}+C)$$
$$= \overline{D}(0+1)$$
$$= \overline{D}$$

Repeat as follows for fig.16.11b.

The Boolean Expression Y $= \overline{B}\mathbf{D}$

CHAPTER-17

Converters

Although digital information processing devices like digital computers and microcomputers are used, information is frequently in analogue form (such as speech, music, and video signals).

This information must first be transformed from its analogue to digital form before being processed using digital procedures. An analog-to-digital converter (A/D converter) is the tool that does this.

Additionally, there are numerous instances where it is required to convert digital information to analogue information since many different types of electronic equipment are intrinsically analogue devices (e.g., stereo amplifiers, radio, and television receivers).

A digital-to-analog converter, often known as a D/A converter, is used to do this.

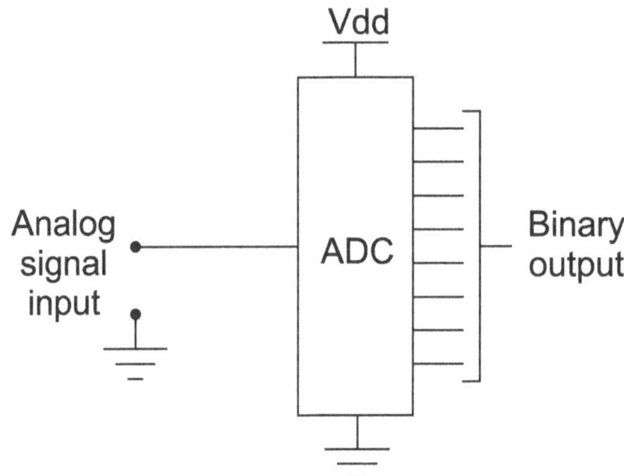

Fig. 17.1 shows an A/D converter's block diagram

A binary number is produced by an ADC from an input analogue electrical signal like voltage or current. While A DAC, which accepts a binary number as input and produces an analogue voltage or current signal as output, can be used to depict it in block diagram form. It resembles fig.17.2 in block diagram form. A/D converters (ADCs), which are quick, precise, and inexpensive, are the result of a considerable lot of technical error.

The importance of DACs cannot be overstated. For instance, video monitors translate digital data produced by computers into analogue signals that are used to focus an electron beam on a particular area of the screen. Though it is challenging to construct a precise DAC in practise, DACs are conceptually easier than ADCs. D/A conversion will be covered before A/D. But first, let's review some fundamental concepts.

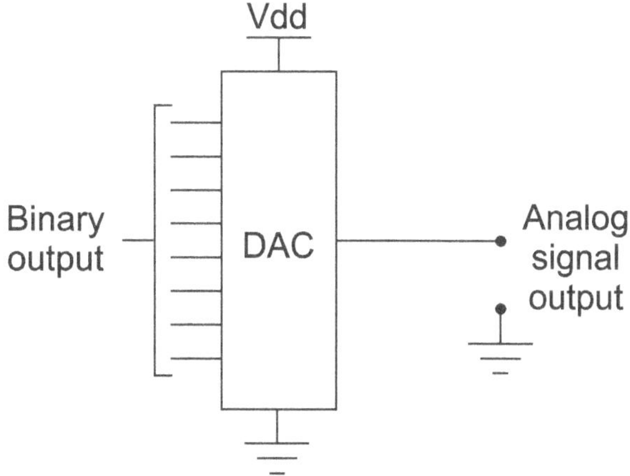

fig. 17.2 Block diagram of D/A converter

They frequently function as a complete interface with analogue sensors and output devices for control systems, like those used in automotive engine controls.

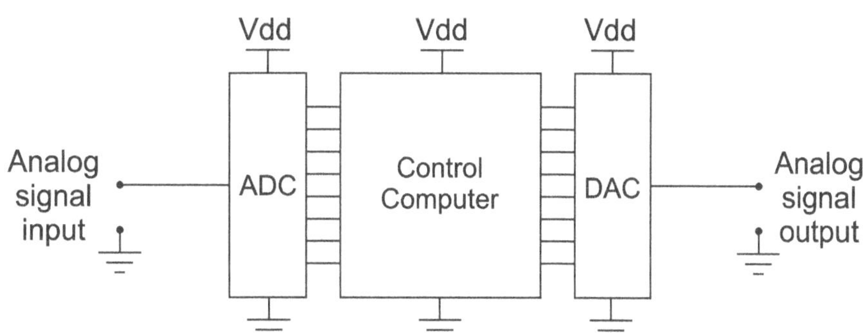

fig. 17.3 Digital control system with analog input/output

17.1 THE CONVERSION OF D/A

The simplest analogue divider—the resistor—is the fundamental component of a DAC. In order to begin, it is necessary to go through the two key characteristics of an operational amplifier connected in an inverting arrangement. Figuring this out is Fig. 17.4.

1. The "-" input is essentially at ground (virtual ground), which is one of the two key characteristics
2. $G = V_{out}/V_{in} = -R_2/R_1$ is the formula for voltage gain. For a current at the input of $I_{in} = V_{in}/R_1$, an equivalent assertion is that the voltage at the output is $V_{out} = GV_{in} = -R_2 I = V_{in} R_2/R_1$.

The trans-conductance, G, is sometimes expressed in the form $Vout = GI_{in}$, where $G = -R_2$ in this instance.

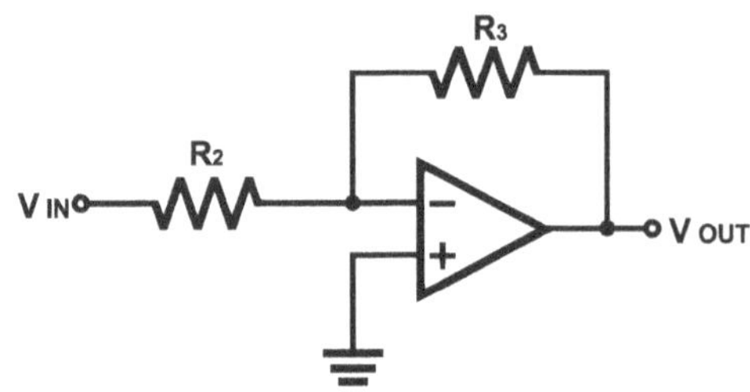

fig. 17.4, Inverting op-amp configuration

The 4-bit example shown in Fig. 17.4 clearly clarifies the fundamental concept of the majority of DACs. The a0 - a3 switches' positions are determined by the input 4-bit digital signal. While a LOW input bit links to ground, a HIGH input bit would be equivalent to a switch connected to 1.0 V. A binary input of 1010, or 1010, is represented by the setup in the figure. There can be no current flow through a switch that is linked to ground because the virtual ground maintains the op-amp input at ground. The current sent to the op-amp, however, will be equal to 1.0 V divided by the resistance of that leg for switches connected to 1.0 V. Next, some current is contributed by all legs with HIGH switches. The desired outcome can be attained by using the binary progression of resistance values depicted in the image.

Example:

I = 1.0/R + 1.0/(4R) = 5/(4R) is the total current to the op-amp. Vout = - RI = 5/4 = 1.25V is the output voltage.

We discover Vout = 15/8 V when all of the input bits are HIGH (1111 = 1510).

A quick check of our strategy reveals that it works as expected (5/4)=(15/8) = 2/3 = 10/15 = 1010=1111.

BinaryDigit	Output voltage
000	0.00 V
001	-1.25 V
010	-2.50 V
011	-3.75 V
100	-5.00 V
101	-6.25 V
110	-7.50 V
111	-8.75 V

For instance, by employing a feedback resistance with a value of 1.6R rather than 2R, we can produce an analogue output that directly corresponds to the binary input (011 = -3 volts, 101 = -5 volts, 111 = -7 volts, etc.) if we choose +5 volts for a "high" voltage level and 0 volts for a "low" voltage level.

17.2 CONVERSION OF A/D

The so-called successive-approximation ADC is one technique for resolving the issues with the digital amp. ADC. This design only differs in a very unique counter circuit called a successive-approximation register. This register counts by testing each value of a bit starting with the MSB and LSB rather than counting up in binary order.

In order to determine whether the binary count is less than or larger than the analogue signal input during the count process, the register continuously checks the comparator's output and modifies the bit values accordingly. The "trial-and-fit" approach of converting decimal to binary is exactly how the register counts. Wherein various bit values, from MSB to LSB, are tested in order to produce a binary number that is equivalent to the original decimal number. Since the DAC output converges on the analogue signal input in significantly greater stages than with the 0-to-full count sequence of a normal counter, this counting approach has the advantage of producing results much more quickly.

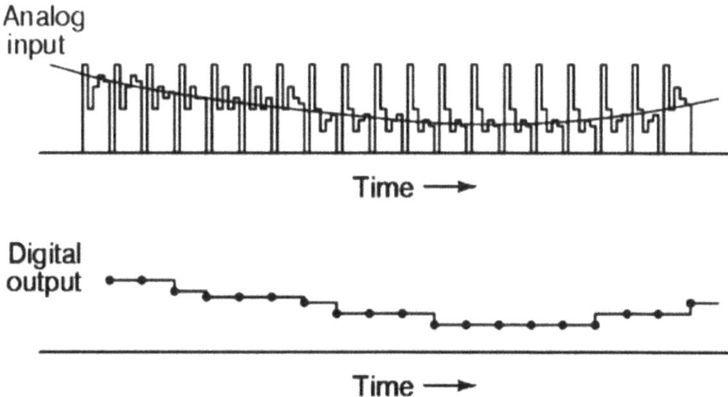

fig. 17.5, Counting ADC

The comparator output is LOW (logical 0) when the binary counter's count is high enough for the DAC output value to be greater than the analogue input value. The counter is no longer receiving new clock pulses, therefore it stops counting. The resulting n-bit digital output is created from the counter's output.

fig. 17.6 SAR ADC with Time

There is no need for a shift register because the successive-approximation register (SAR) can typically output the binary number in serial (one bit at a time) format. The use of a subsequent device is shown on a timeline as-The register continuously checks the comparator's output during the count operation to see if the approximate ADC resembles Fig. 17.6.

17.3 R-2R LADDER

Although it is a rather small detail, this represents an intriguing concept. Due to the fact that it only utilises two resistor values, the "R-2R ladder" is of practical interest. This is advantageous since it is challenging to precisely produce resistors with arbitrary resistance.

The R-2R's two resistances are to be contrasted with the circuit in Fig. 17.7, which uses as many different resistance values as there are bits. Comprehending the pattern of equivalences shown in Fig. 17.8, which may be used to reproduce an indefinitely long ladder and hence handle in an arbitrary amount of bits, is essential to comprehending the R-2R ladder.

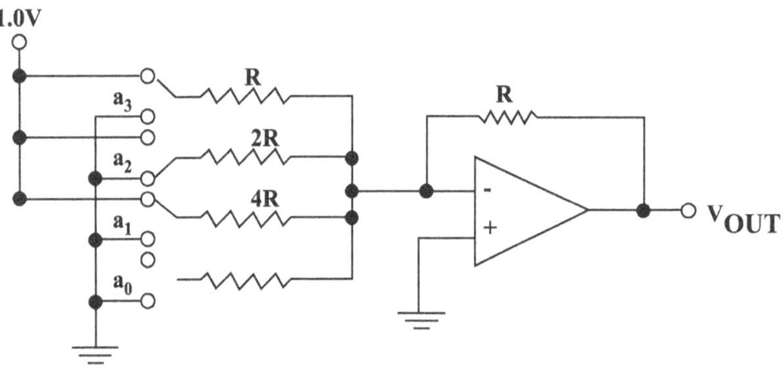

fig. 17.7 4-bit DAC scheme

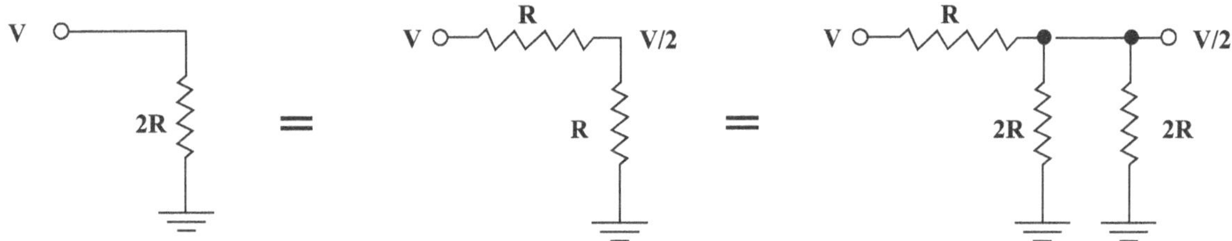

fig. 17.8 R-2R ladder with equivalent circuit.

CHAPTER-18

Memory
Terminology

Data storage technologies as a subject will not be covered here. The issue of how data storage can be deliberately organised is what most interests us in this situation. The fact that the memories we will examine are RAM—random access memories—is an essential aspect in common with them.

With a valid input address, each piece of information can therefore be stored or retrieved independently. When compared to sequential memories, like those used for magnetic tape storage of data, where bits must be saved or retrieved in a specific order, this is the opposite. Unfortunately, the definition of RAM has changed, now referring to a memory that allows for simple bit storage, retrieval, writing to, and reading from.

The bit is the tiniest piece of digital data. The binary digit used to represent it can either be a 1 or a 0. A nibble is a grouping of four bits, whereas a byte is a grouping of eight bits. A data word is a collection of 16 bits.

four bits per nibble (1111)

(8 bits per byte) (11111111)

16-bit data word (1111111111111111)

A megabit (MB) is 1K X 1K bits, while a "K" of memory is $2^{10} = 1024$ bits (sometimes abbreviated KB).

Data can be stored in any of the aforementioned ways, depending on the size of the memory System. No matter how many bits are grouped together, every grouping has a unique address. Two numbers are typically used to describe memory systems. The bit size is the first number, and the number after that represents the number of addressable spots.

Instance: 16 x 256K

There would be 256,000 addressable places in this memory system, with each addressable location containing 16 bits.

The number of bits in the address bus determines how many addressable places are available. The address bus would require 16 address lines for a 256K memory. We could address $2^{16} = 262,144$ sites with 16 address lines. Then, 16 bits might be stored in each addressable position.

The bistable multi vibrator, as depicted in fig. 18.1, is a very basic form of electronic memory. It is volatile (needs electricity to retain its memory) and extremely quick, with the capacity to store a single bit of data. With the D input acting as the data "write" input, the Q output as the data "read" output, and the enable input acting as the read/write control line, the D-latch is likely the most straightforward implementation of a bi stable multi vibrator for memory utilisation.

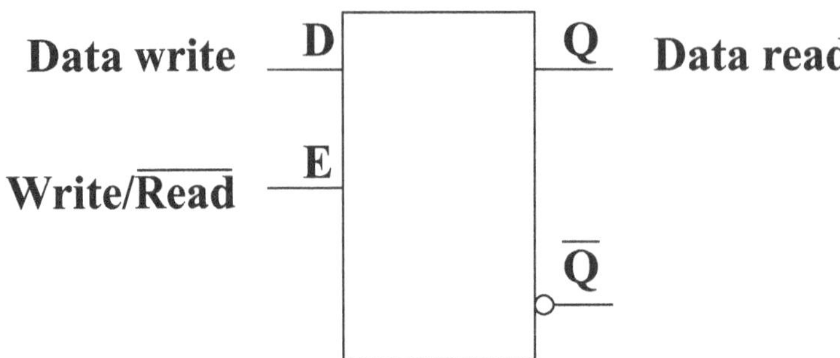

fig.18.1bistable multi vibrator

We will need a lot of latches set up in some sort of array so that we can choose which latch (or sets of latches) we want to read from or write to if we want more storage than one bit, which we probably do. By connecting a common data bus line to a pair of tri state buffers, we can enable those buffers to connect the Q output to the data line (READ), the D input to the data line (WRITE), or to remain in the High-Z state to keep D and Q disconnected from the data line (unaddressed mode). Internally, one memory "cell" would resemble fig. 18.2.

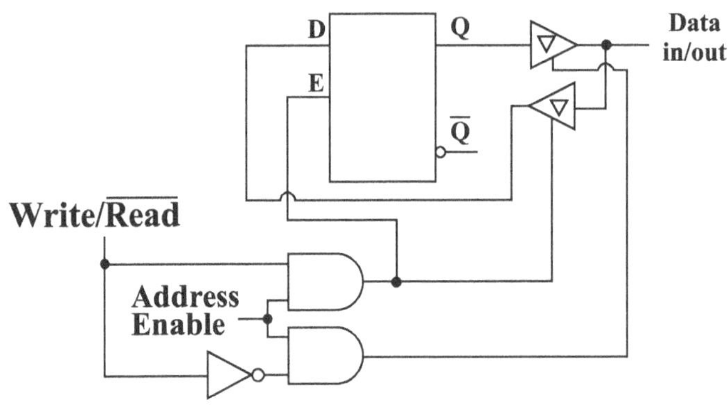

fig.18.2Memory cell circuit

18.1 VARIOUS TYPES OF MEMORIES

Basically, there are two types of memories:

a) Random Access Memory-RAM
b) Read-only memory-ROM

18.1.1 RAM stands for random access memory.

Volatile or non-volatile memory integrated circuits are two different types. When the power to the IC is switched off, the data in a volatile memory is lost, whereas the data in a non-volatile memory is retained. Data is lost when the IC's power is switched off, which classifies Random Access Memory (RAM) as volatile memory. RAM memory ICs can be divided into two categories.

i) Static RAM (SRAM)
ii) Dynamic RAM (DRAM)

18.1.1.1   SRAM, or static RAM

Data that doesn't need to be refreshed is stored in static RAM. These memory units store information in latch form. Figure 18.3 depicts a sample SRAM data storage device. These are essentially flip-flop arrays. They can be produced in integrated circuits (ICs) as huge arrays of colour flip-flops.SRAM is fundamentally a little quicker than dynamic RAM. This straightforward memory circuit is volatile and random-access. It is referred to as a static RAM technically.

It has a 16-bit overall memory capacity. It would be referred to as a 16 x 1 bit static RAM circuit since it has a data bus that is 1 bit wide and 16 addresses. As you can see, building a functional static RAM circuit requires a huge number of gates (and many transistors per gate!). As a result, the static RAM has a lower capacity per unit IC chip space than the majority of other types of RAM technology, making it a relatively low-density device.

Because each cell circuit uses a specific amount of power, the total power consumption for a wide array of cells can be rather significant. Early static RAM banks in personal computers used a lot of electricity and produced a lot of heat. Low storage density is still a problem, despite the fact that CMOS IC technology has made it feasible to reduce the particular power consumption of static RAM circuits.

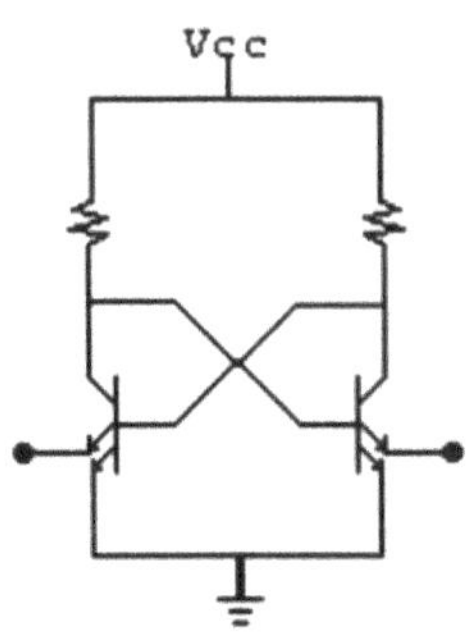

fig.18.3 Static memory cell

18.1.1.2   DRAM ,Dynamic RAM

Arrays of capacitors are used in dynamic RAM (DRAM). If a capacitor is charged and its voltage $V = Q/C$ is higher than the required level for the current logic standard, a HIGH bit is generated. The term "dynamic" refers to information that must constantly be updated since the charge will leak out through the connections' resistance in intervals of order 1 msec. Static RAM cannot be manufactured with as many bits per unit area as dynamic RAM. So, for the majority of large-scale IC memory, it is typically the technology of choice.

The fact that DRAM memory systems require extra circuitry to keep their data current is one of its disadvantages. This is due to the fact that a DRAM system stores data in a capacitor-type of device, which requires frequent recharge as the capacitor starts to drain and whenever data is retrieved from it. Figure 18.4 illustrates a DRAM cell as an example.

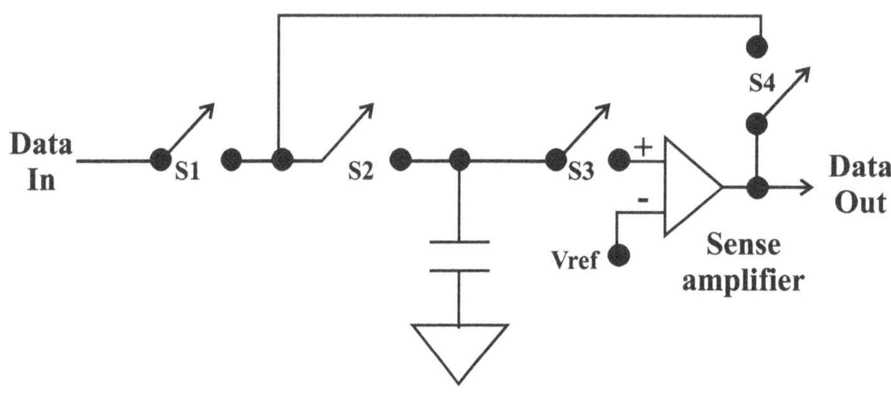

fig. 18.4 DRAM cell

The address decoding circuitry closes switches S1 & S2 while leaving switches S3 & S4 open when data is stored in the cell. Switches S1 is opened while switches S2, S3, and S4 are kept closed when data is read from the cell. By providing a feedback path through switch S4 to reload or recharge the capacitor, this enables data to be read from the sense amplifier.

Memory access time is a DRAM system's principal drawback when compared to an SRAM system. The DRAM system's configuration results in a longer time required to retrieve data from its memory structure. The SRAM system's disadvantage is that the circuit is larger and hence more expensive.

18.2 READ ONLY MEMORY (ROM)

The design of read-only memory (ROM) is similar to static or dynamic RAM circuits, with the exception that the "latching" mechanism in ROM is intended for single (or limited) operations. As an illustration, it is typically wise to save the instructions necessary to initialise a computer when first powering it on in ROM. To represent the two binary states, the simplest sort of ROM uses microscopic "fuses" that can be deliberately blown or left alone.

Such ROM circuits can only be written once since, as is obvious, once one of the tiny fuses blows, it cannot be repaired. These circuits are commonly referred to as PROMs (Programmable Read-Only Memory) since they can only be written (programmed) once.

Not all writing techniques, meanwhile, are as irreversible as blown fuses. A memory device that resembles a cross between a RAM and a ROM can be created if a transistor latch that can only be reset with great difficulty can be created. The name of such a gadget, EPROM (Erasable Programmable Read-Only Memory), is somewhat oxymoronic.

Electrically-erasable (EEPROM) and ultraviolet-erasable (UV/EPROM) EPROMs are the two main types. Both kinds of EPROMs latch on or off using capacitive charge MOSFETs. UV/EPROMs are "cleared" through prolonged ultraviolet light exposure. They are simple to spot since they have a transparent glass window that lets light shine through the silicon chip substance. You must tape over that glass pane after programming it to keep incoming light from deteriorating the data over time. Higher signal voltages than those used in "read-only" mode are frequently utilised to programming EPROMs.

18.3 READ AND WRITE MEMORY

A write operation is the act of placing data in memory. To accomplish this, the data to be saved is placed on the data bus, and the address to which the data is being sent is placed on the address bus. Data is stored in memory as soon as the memory device is enabled. Figure 18.5 illustrates a store data link between a central processing unit (CPU) and a memory device. Keep in mind that the arrows indicate the flow of information.

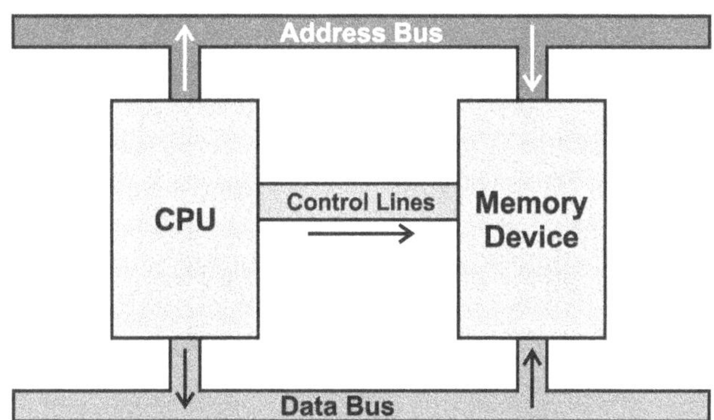

fig. 18.5 Data write operation

A read operation is the process of obtaining data from memory. In order to accomplish this, a controlling device must send a request for data and place the data's address on the address bus. The memory device then accesses the data associated with that specific address and transfers it to the data bus. Figure 18.6 provides an illustration of a controlling device reading data from memory.

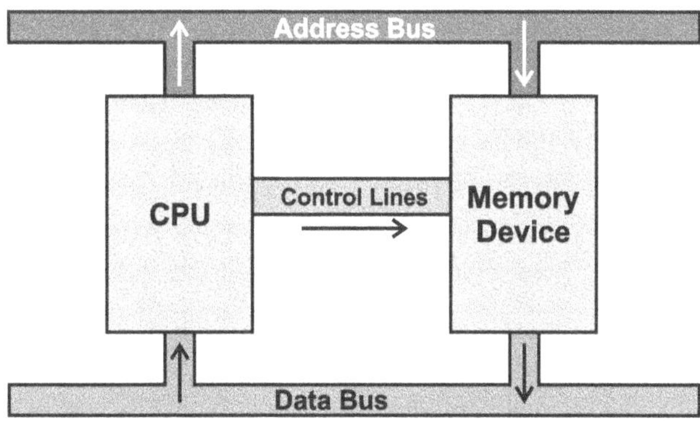

fig. 18.6 Data read operation

18.4 R/W 4-BIT DECODER

A 4-bit binary code fed into the decoder depicted in fig. 18.7 is used to address each of the 16 memory cells independently. A cell's internal tri state buffers will cut it off from the 1-bit data bus if it is not addressed, making it impossible to write to or read from that cell through the bus. The data bus will only provide access to the cell circuit that is targeted by the 4-bit decoder input.

16 x 1 bit Memory

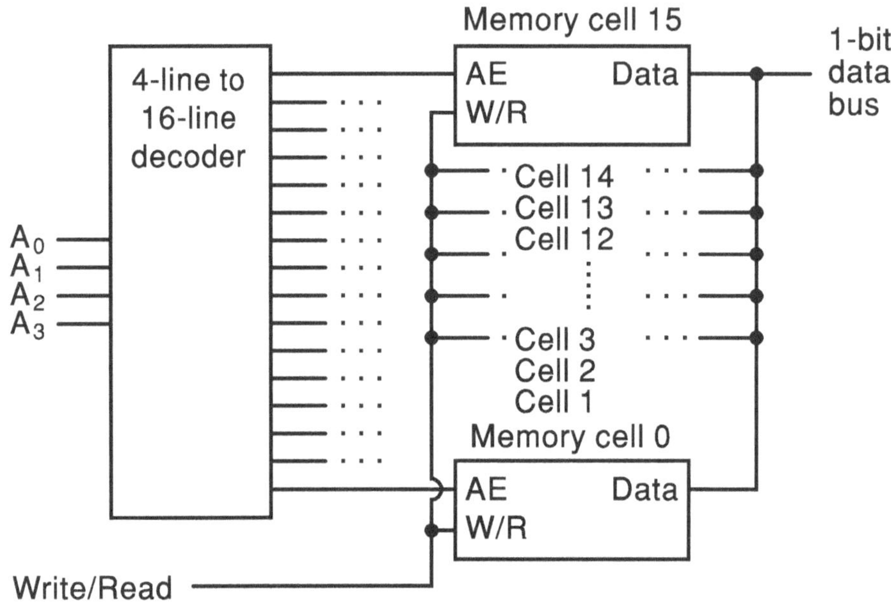

fig. 18.7 16 X 1bit memory cell

This straightforward random-access, volatile memory circuit. It is known as a static RAM in technical terms. It has 16 bits of total memory. It would be referred to as a 16 x 1 bit static RAM circuit since it has 16 addresses and a data bus that is 1 bit wide. As you can see, putting together a functional static RAM circuit requires a staggering amount of gates (and many transistors per gate!). As a result, the static RAM has a lower capacity per unit IC chip space than the majority of other types of RAM technology.

The total power consumption for a big array of cells can be rather significant because each cell circuit uses a specific amount of electricity. In the beginning, static RAM banks in personal computers used a lot of power and produced a lot of heat. Low storage density is still a problem despite the fact that CMOS IC technology has made it feasible to reduce the specific power consumption of static RAM circuits.

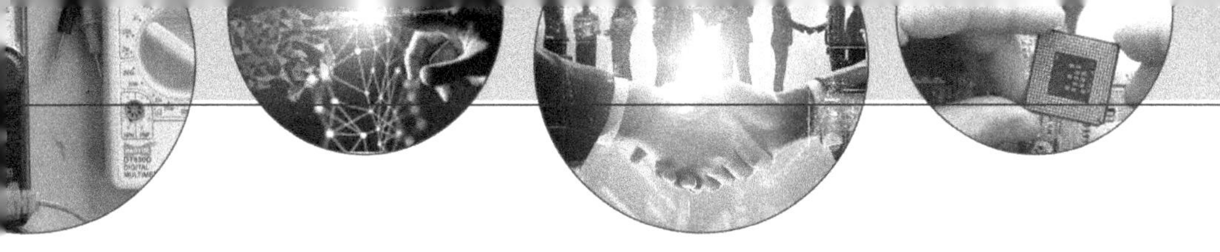

blank

PART- III

MICROPROCESSOR

CHAPTER-19

Introduction to Microprocessor

19.1 SYSTEM WITH MICROPROCESSORS

The logical function of a microprocessor, a Very Large Scale Integrated (VLSI) logic integrated circuit, is managed by an instruction code. A microprocessor needs this instruction code, which is a particular arrangement of logic bits, to carry out its logic operations. Instructions to carry out logical, mathematical, or input/output tasks may be included in these codes.

One kind of computer system is one that uses a microprocessor. A processing unit, memory, and input/output are the three fundamental functional building blocks of all computer systems. In reality, the address bus, data bus, or control lines are used to link these parts together.

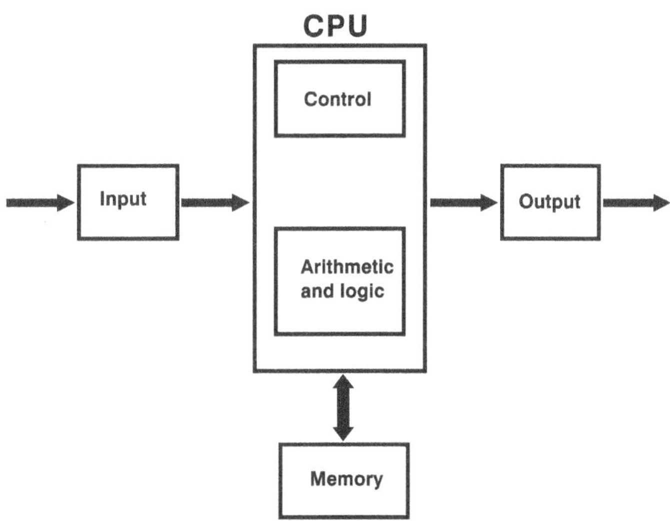

fig. 19.1 Basic Microprocessor System

Fig. 19.1 gives a case in point for this. The central processing unit (CPU) manages the timing and functioning of all other microprocessor operations in addition to carrying out the computational duties for the microprocessor system.

Recall: This is normal Random Access Memory (RAM) or Read-Only Memory (ROM)

Units of input/output: These gadgets transmit and receive data from external devices.

All of the microprocessor's address lines are located on the address bus.

The address bus connects all external memory and I/O devices that require addressing to the relevant microprocessor address lines. The address bus only goes one way.

Data Bus: The data bus is where all information is sent to and received from the CPU. Keep in mind that the data bus needs to be bidirectional.

A computer must be able to recall its actions as well as information from inputs like sensors or the keyboard. For this purpose, memory is used. When a piece of data (a "Byte") has to be used, the processor "fetches" it from its address by storing it in a memory location called a "ADDRESS." LOCATION is another term for the ADDRESS.

It's critical to comprehend how data is transferred between computer components. The memory is often where data is kept. 'BITs' are the unit of storage for data, which can be numbers, words, or characters. These represent "1s" and "0s." These bits are organised into "BITS," which are a series of 1s and 0s.

A letter from the alphabet or a straightforward instruction for the computer could be represented by the eight BITS in the table below. Since words and sentences are not understood by computers, they must first be converted into BITs and BYTES before the computer can use them. Consider an eight bit byte as an example; how do you imagine a sixteen bit byte to actually look? An address or location in memory is assigned to each byte. Eight bits, or one byte, are equal to each 1 or 0.

1	0	0	1	1	0	1	1

19.2 MICROPROCESSOR

The traditional definition of a "microprocessor" is a single "chip." An example is a computer's CPU (Central Processing Unit), which manages the majority of a computer's operations. But technically speaking, a microprocessor is just a group of chips. Calculator is a pretty nice example. Microprocessors are used to operate many equipment, including washing machines. The microprocessor manages every part of the washing cycle, including the temperature and wash type. What additional equipment or gadgets are controlled by microprocessors?

The word size of the CPU was reduced from 32 bits to 4 bits to allow the transistors of its logic circuits to fit on a single component, giving rise to the microprocessor. The central processing unit (CPU) in a computer system, embedded system, or portable device is typically one or more microprocessors. In the middle of the 1970s, microprocessors enabled the development of the microcomputer. Prior to this time, electronic CPUs were primarily constructed from large discrete switching devices (and later, small-scale integrated circuits) that had a limited number of transistors.

The cost of CPU power was significantly decreased by integrating the processor onto one or a very small number of large-scale integrated circuit packages (carrying the equivalent of thousands or millions of discrete transistors). The microprocessor has replaced all other types of the CPU almost entirely since the introduction of the IC in the middle of the 1970s.

When discussing microprocessors and computers, confusion is common. A microprocessor, like a calculator, is made up of a number of chips (Integrated Circuits). Microprocessors are among the chips that make up a computer.

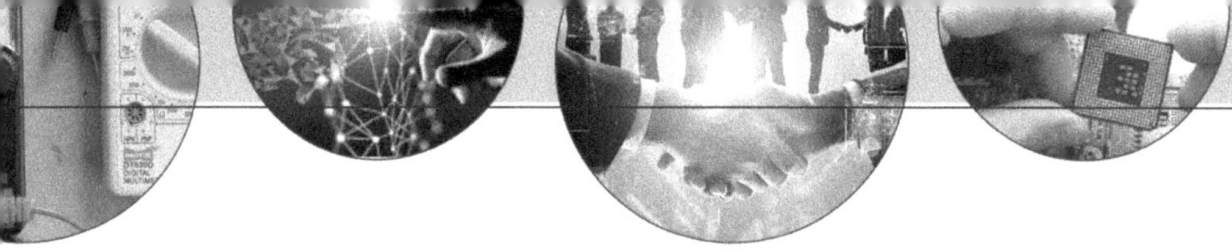

The digital computer's microprocessor is a component. As shown in the table in fig. 19.3, it can be further separated into:

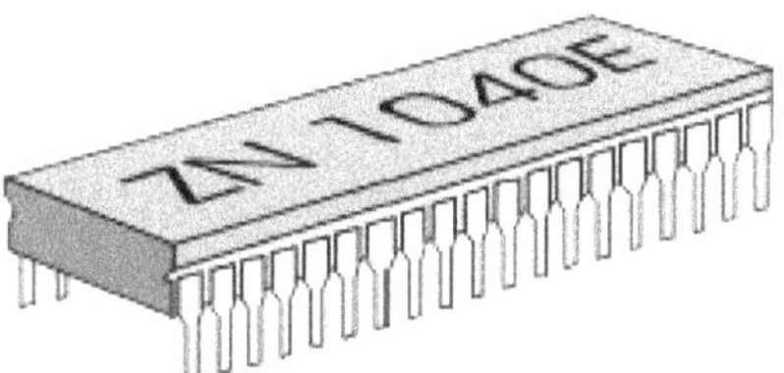

fig. 19.2 Microprocessor Chip

1. ALU   2. A Control unit   3. A Program counter   4. A Register unit

ALU, or arithmetic logic unit:carries out all Boolean operations and calculations in mathematics.

Control Unit : Enables data flow to and from the ALU, internal registers, and external devices. Decodes instructions read from programmes in memory. All devices sharing an address and/or data bus share a control bus, which arbitrates all bus communications.

The Program counter : Sends an incremental address value to any external memory chip(s) in order to retrieve the subsequent instruction. Memory connected by an address bus.

A Registers Unit : Small read/write memory with the "accumulator" that carry data for ALUs, error codes, instruction codes, etc. Data Bus from external devices and RAM memory.

fig. 19.3 Block diagramc of Microprocessor

In much the same way that the human brain adapts the strength of inter-neural connections depending on environmental experiences, which explains why memory retention improves with repeated study and behaviour is modified through consequential feedback, engineers and computer scientists looked forward to the possibility of creating digital devices that could modify their own programming. This would only be feasible if the data and computer program were kept in the same "pool" of writable memory.

It's fascinating to notice that computer scientists still considers the idea of a self-modifying programme to be cutting edge. The majority of computer programming uses relatively stable sequences of instructions, with the only data that is changed being a different field of data.

Together, read/write memory, an ALU, and a control unit form a digital device that is frequently referred to as a processor. It is referred to as a microprocessor if little memory is utilised and all the required parts are located on a single integrated circuit. It is referred to as a central processing unit, or CPU, when coupled with the required bus-control support circuitry.

The main memory's address lines get this binary address value from the control unit, which then reads the memory's data output to transfer it to another holding register. The Control Unit correctly addresses the location of the requested data and directs the data output to ALU registers if the fetched instruction necessitates reading additional data from memory. For instance, when adding two numbers, we must read both of the numbers from main memory or from another source. The Control Unit would then carry out the order by informing the ALU to perform the specified action with the two integers and sending the outcome to an additional register known as the accumulator. The Control Unit now advances the programme counter to step the following instruction once the instruction has been "fetched" and "executed," and the cycle repeats.

In some microprocessor designs, there are small programmes (known as microcode) that handle all the sub-steps required to perform more sophisticated math operations that are stored in a specific ROM memory inside the device. The programmer won't have to worry about attempting to explain to the microprocessor how to carry out each intricate step because only one instruction will need to be read from the programme RAM in this case.

19.3 MICROPROCESSOR'S EVOLUTION

When it comes to consistently improving performance over time, Moore's Law has been seen to be followed by the development of microprocessors. According to this law, an integrated circuit's complexity should double every 18 months in terms of the cost of the cheapest component. Since the early 1970s, this adage has generally shown to be true. Microprocessors are currently used in every type of computer system, from the largest mainframes to the smallest handheld computers, thanks to their constant advancement in power from their humble beginnings as the calculator's drivers.

A complete microprocessor may have been created at roughly the same time by the Central Air Data Computer (CDAC), a project by Garrett Ai Research, the Texas Instruments (TI) TMS 1000, and the Intel 4004.

The US Navy commissioned Garrett Ai Research to create a digital computer in 1968 to compete with electromechanical systems that were being developed at the time for the new F-14 Tomcat fighter's primary flight control computer. The team behind the project included designer Ray Holt and Steve Geller. By 1970, the plan was finished, and the central processing unit was a chipset based on MOS. The design differed greatly from early Tomcat versions and was (about 20 times) smaller and significantly more dependable than the mechanical systems it competed against.

This system contained "A 20-bit, pipelined, parallel multi-microprocessor". However, the Navy did not permit publishing of the concept until 1997 because they believed the system to be too sophisticated. Due to this, the CADC and the MP944 chipset it utilised are still largely obscure. a first microprocessor chip set (see below). The 4-bit TMS 1000 was created by TI, who also emphasised pre-programmed integrated. On September 17, 1971, the TMS1802NC, a version that had a calculator on a chip, was introduced. The Intel chip was the 4-bit 4004, created by Federico Faggin and Marcian Hoff, and it was launched on November 15, 1971.

A microprocessor variant known as a "computer-on-a-chip" combines the CPU, some memory, and I/O (input/output) lines on a single chip. Gary Boone and Michael J. Cochran of TI were given the U.S. Patent 4,074,351 for the computer-on-a-chip invention, which was referred to as the "microcomputer patent" at the time. Aside from this patent, a "microcomputer" is a computer that has one or more microprocessors acting as its central processing unit (CPU), whereas the idea described in the patent may be closer to a "microcontroller."

According to MIT Press's A History of Modern Computing, pp. 220–221, Intel signed a contract for a chip with Computer Terminals Corporation, afterwards known as Data Point, of San Antonio, Texas, for a terminal they were creating. Data Point later decided against using the device, and in April 1972 Intel began selling it under the 8008 brand. This was the first 8-bit microprocessor ever created.

Following the 4004 in 1972, the well-known "Mark-8" computer kit was introduced and featured the 8008, the first 8-bit microprocessor ever created. These CPUs served as the foundation for the wildly popular Intel 8080 (1974), Zilog Z80 (1976), and subsequent Intel 8-bit processors. In August 1974, the Motorola 6800, a rival model, was released. In 1975, the MOS Technology 6502 used a clone of its architecture that had been modified, and its popularity in the 1980s rivalled that of the Z80.

- The National Semiconductor IMP- 16 was the first multi-chip 16-bit microprocessor, and it was unveiled in early 1973
- The MC68000, which debuted in 1979, is the most notable of the 32-bit designs
- Although 64-bit microprocessor designs have been in use in a number of applications since the early 1990s, 64-bit PC-specific microchips have just recently been made available.

Speed and computational power improvements come along with an increase in bit count. The quantity of data that can be input into or exported from the CPU is determined by these bits.

Year	Company	Processor
1971	Intel	4004 (4 bit)
1974	Intel	8080 (8 bit)
1975	Motorola	6800 (8 bit)

fig. 19.4 Evolution of Microprocessor

19.4 PROGRAMMING FOR MICROPROCESSOR

The "language" of the CPU is used when a human programmer creates a set of instructions to directly advise a microprocessor on how to carry out a task (such as automatically controlling the rate of fuel injection to an engine). The term "machine language" is frequently used to describe this language, which is made up of the exact same binary codes that the Control Unit inside the CPU chip decodes to carry out tasks.

Although machine language programs can be written in binary notation, hexadecimal is more frequently used since it is easier for readers to understand. I'll only use a few of the widely used Intel 8085 microprocessor instruction codes as examples.

Hexadecimal	Binary	Instruction description
7B	01111011	Transfer register A's contents to register E.
87	0000111	Add contents of register A to register D
1C	00011100	Register E's contents are increased by 1
D3	11010011	Output data to the data bus as a byte

These instructions can be misunderstood and forgotten even with hexadecimal notation. Another programming tool for this purpose is assembly language. When describing programme steps in assembly language, two to four letter mnemonic terms rather than the actual hex or binary code are employed. For instance, the assembly language representation of instruction 7B for the Intel 8080 would be "MOV A,E". The mnemonics are obviously useless to a microprocessor because they can only read binary codes, but they do provide programmers with a quick way to manage creating their programmes on paper or in a text editor (word processor). To avoid ever having to deal with strange hex or tedious binary code notation, there are even programmes created specifically for computers called assemblers that comprehend these mnemonics and translate them to the appropriate binary codes for a given target microprocessor.

A programme created by a person must first be stored in memory before a microprocessor can run it. If the programme needs to be stored in ROM (which some do), this can be done using a ROM programmer or, if you're feeling particularly masochistic, by placing the ROM chip on a breadboard, connecting the correct wires to the address and data lines one at a time for each instruction, and powering the device up.

If the programme needs to be entered manually, even if the computer is too stupid to do anything else, it might be possible to do so using the keyboard of the operating system's computer (some computers have a mini-program stored in ROM that instructs the microprocessor how to accept keystrokes from a keyboard and store them as commands in RAM). Such is the operation of many "hobby" computer kits. If the computer being written is a fully complete personal computer with an operating system, disc drives, and everything else, you can just tell the assembler to save your final programme to a disc for later retrieval.

The microprocessor's Programme Counter register would be set to point to the position ("address") on the disc where the first instruction is recorded, and your programme would start from there. To "run" your programme, you would simply write its filename at the prompt and press the Enter key. Although writing code in machine language or assembly language produces quick and highly effective programmes, doing so requires a lot of effort and expertise for all but the most straightforward jobs because each machine language instruction is so rudimentary.

Making it possible for programmers to write in "high level" languages, which are better at expressing human cognition, is the solution to this problem Instead of typing hundreds of complicated assembly language commands, a programmer could create something like this using a high-level programming language. Without providing the computer with any other instructions, simply expect it to print "Hello, IICT!" Despite the fact that this is a great idea, how can a microprocessor, with its confined language, understand such "human" thinking?

Although interpretation is straightforward, a slow-running program results from the microprocessor's constant translation of the program between steps, which takes time. The initial translation of the entire program into machine code requires time during compilation, but the final machine code requires no further translation and runs faster as a result. Interpreted programming languages include FORTH and BASIC. Compilation is used for languages like C, C++, FORTRAN, and PASCAL. Because of the effectiveness of the finished product, compiled languages are typically regarded as the languages of preference for professional programmers.

CHAPTER-20

Microprocessor &
Its Architecture

The foundation and fundamental ideas common to microprocessors were laid out in the preceding chapter. Midway through the 1970s, Intel produced the 8085, an 8-bit microprocessor. Its binary compatibility with the more well-known Intel 8080 allowed for the construction of smaller and more affordable microcomputer systems because it needed less supporting hardware.

The 8085 only needed a +5-volt (V) power supply, as opposed to the +5V, -5V, and +12V supplies the 8080 required, hence the "5" in the model number. The 8085 subsequently found use as a micro controller (mostly because of its ability to lower component count), and both CPUs were occasionally utilized in computers running the CP/M operating system. The compatible but superior Zilog Z80, which dominated the majority of the CP/M computer market and a sizable portion of the expanding home computer market in the early to mid-1980s, superseded both designs for desktop computers. As a controller, the 8085 was used for a very long time. It continued to be used for new production throughout the lifespan of those devices (which was often far longer than the life cycle of desktop computers) after being In the late 1970s, research was put into products like the VT100 video terminal and the DEC tape controller.

20.1   MICROPROCESSOR ARCHITECTURE (8085)

Figure 20.1 illustrates the 8085 Architecture, which has an 8-bit data bus and a 16-bit address bus. The 8085 increased the level of system integration by incorporating all of the timing and control, serial I/O control functions. There are 10 different registers located inside the 8085. A, B, C, D, E, H, L, PSW, PC, and SP are their names. The PSW, PC, and SP registers are the only ones not being used for temporary storing of data required by the programme. Another distinction between the A accumulator and the other registers is its name. It is utilised to tally the outcomes of numerous commands, such as add and sub (subtract). The Stack Pointer (SP), which is 16 bits wide and actually carries addresses, is different from the Programme Counter (PC), which we have already explained. The width of the others is 8 bits.

Control Unit (CU)

It creates signals inside the CPU to execute the decoded instruction. Actually makes certain connections between UP blocks open or closed, allowing data to travel where it is needed and enabling ALU processes.

ALU, or Arithmetic Logic Unit

The real mathematical and logical operations, such as "add," "subtract," "AND," "OR," etc., are carried out by the ALU. does arithmetic using information from the Accumulator and from memory. Accumulator always stores operation results.

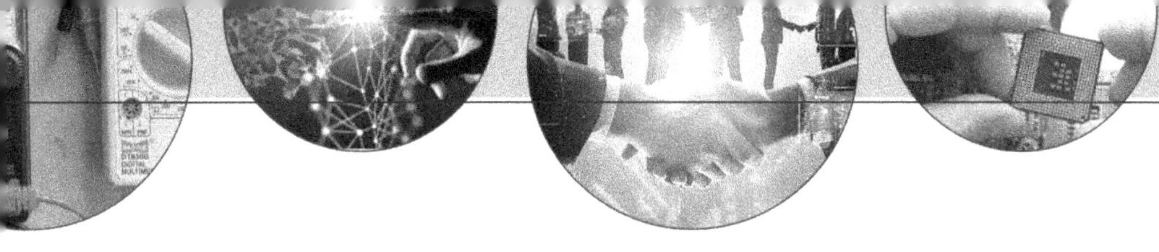

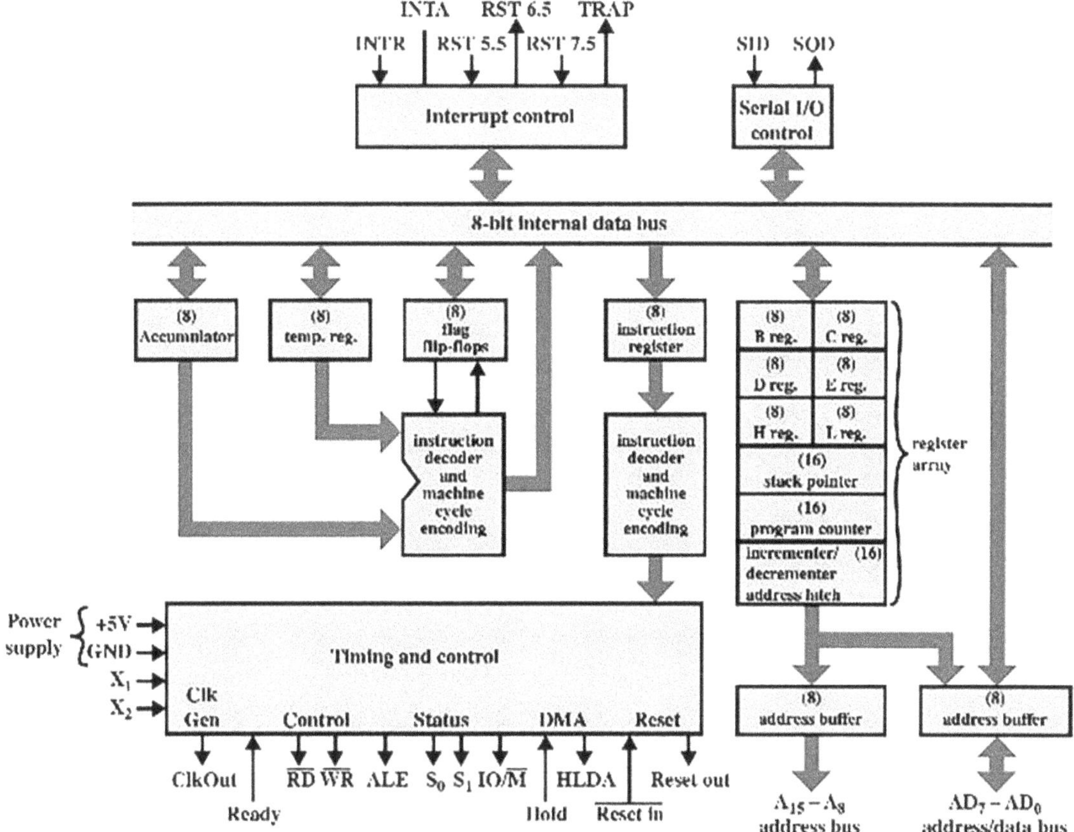

fig. 20.1 Internal architecture of 8085 microprocessor

Registers

According to Fig. 20.1 there are six registers, an accumulator, and a flag register in the 8085/8080A programming model. It also contains two 16-bit registers, the programme counter and the stack pointer. Six general-purpose registers, labeled B, C, D, E, H, and L in the illustration, are available on the 8085/8080A to hold 8-bit data. To carry out various 16-bit operations, the register pairs BC, DE, and HL can be merged. By employing data copy instructions,

These registers can be used by the programmer to copy or save data. The 8085 is capable of accessing 216 (= 65,536) separate 8-bit memories.

Its address space, or alternatively, is 64 KB. It includes a separate address space for up to 28 (=256) I/O ports, unlike several other microprocessors of the time. The CPU features four interrupt types: three maskable (RST 7.5, 6.5, and 5.5), one non-maskable (TRAP), and one externally serviced (INTR). The RST n.5 interrupts make use of actual processor pins, allowing basic systems to do away with the expense of a separate interrupt controller hardware.

Accumulator

An 8-bit register called the accumulator is a component of the ALU, or arithmetic/logic unit. In addition to storing and retrieving 8-bit data, this register is utilised to carry out arithmetic and logical operations. An operation's result is kept in the accumulator. The register A is another name for the accumulator.

Flags

The accumulator and other registers' data conditions determine whether the five flip-flops in the ALU are set or reset following an operation. These flags, which go by the names Zero (Z), Carry (CY), Sign (S), Parity (P), and Auxiliary Carry (AC), are enumerated in the Table, and the following Figure shows their bit positions in the flag register. The three flags Zero, Carry, and Sign are the most often employed. These flags are tested by the microprocessor to determine data conditions.

The Carry flag (CY) is set to one by the flip-flop to signal a carry, for instance, if the sum of the two values added exceeds eight bits in the accumulator. The flip-flop designated as the Zero(Z) flag is set to one whenever an arithmetic operation yields a result of zero. An 8-bit register known as the flag register is seen next to the accumulator in Fig. 20.1. Five of the eight bit places are not used as registers; instead, they are used to store the outputs of the five flip-flops. The flags are kept in the 8-bit register so that the programmer can access the register using an instruction to check the flags (data conditions). The microprocessor's decision-making process heavily relies on these flags. Through the programme instructions, the flags' conditions (set or reset) are tested. For instance, when the CY flag is set, the instruction JC (Jump on Carry) is used to alter the program's sequence. Writing assembly language programmes requires a solid understanding of flag.

Programme Counter-PC

This 16-bit register manages the order in which instructions are executed. A memory pointer is contained in this register. This is a 16-bit register since memory locations have 16-bit addresses. This register is used by the microprocessor to organise how the instructions are executed. The programme counter's job is to indicate the memory location where the subsequent byte should be fetched from. The programme counter is raised by one to indicate the following memory address whenever a byte of machine code is fetched.

A STACK POINTER-SP

Additionally employed as a memory pointer, the stack pointer is a 16-bit register. It points to a place in R/W memory known as the stack. By loading a 16-bit address into the stack pointer, the stack's beginning is shown. The chapter "Stack and Subroutines" provides an explanation of the stack idea.

The Decoder or Instruction Register

A short-term storage location for a program's active instructions. the most recent memory-sent instruction sent here before execution. The decoder then accepts the command

and "decodes" or understands it. Instructions that have been decoded are subsequently moved on to the following level.

A memory address register

It stores the next programme instruction's address when it is received from the computer. addresses for the locations where the programme is being run are fed into the address bus.

A Control Generator

Signals are generated by a control generator within the Microprocessor to carry out the decoded instruction. In reality, it is what opens or closes specific connections between uP blocks, allowing data to get to where it is needed and triggering ALU operations.

The Register Selector

The example's register stack is used under the control of this block. The only part of the set that will receive instructions from the control unit is a logic circuit that switches between the various registers.

Registers for other Purposes

Extra registers are needed by the microprocessor for adaptability. can be applied to store extra data throughout a programme. Various registers with various names may be present in more complicated processors.

20.2 SYSTEMS BUSES

The typical system employs a number of cables called buses to transfer binary data, one bit per wire. The Address Bus, Data Bus, and Control Bus are the three buses that a typical microprocessor uses to connect to memory and other devices (input and output). A 16-line address bus that may access 216 memory locations (or 64 KB) in memory. Data bus: An 8-line bus that can access one 8-bit byte of data at a time. As contrast to address bus width, data bus width has traditionally been used to measure processor bit designations, leading to the designation of an 8-bit microprocessor. Control buses: These transport the vital signals for a variety of processes..

Address Bus

There are 16 wires total because there is one wire for every bit. Memory is alerted by the binary number carried to "open" the specified box. Then, binary data can be added to or removed. There are 16 wires in the Address Bus, or 16 bits. It has a 16-bit "width". A 16-bit binary number can contain 216 different numbers, or 32000 different numbers, ranging from 0000000000000000 to 1111111111111111. The amount of memory that may be used depends on the size of the address bus because memory is made up of boxes, each of which has a distinct address. The CPU transmits an address to the memory on the address bus, for example, 0000000000000011 (3 in decimal), in order to communicate with it. The memory chooses a box number.3 is required to read or write data. Data on the address bus can only be sent from the microprocessor to the memory in one direction.

Data Bus

The Data Bus transports binary-coded "data" between the CPU and other external components, like the memory. 8 or 16 bits are typically the size. Performance of the microprocessor is influenced by size, which is defined by the size of the memory boxes and the microprocessor itself. Typically, the Data Bus has 8 wires. So there are 28 different binary digit combinations. Between memory and the microprocessor, "data," or information, mathematical solutions, etc., is transmitted via a data bus.

The bus has two directions. What math can be performed depends on the size of the data bus. The greatest number is 11111111 (255 in decimal) if there are only 8 bits available. Therefore, larger numbers must be divided into 255-bit chunks. This makes the processor slower. The microprocessor receives instructions from memory via the data bus as well. As a result, the size of the bus restricts the number of available instructions to 256, each of which is identified by a unique number.

The Control Bus

Multiple connections known as the "Control Bus" are used to coordinate and control microprocessor operations. Read/write binary digit, one line. Control the process of writing to (storing data in memory) or reading from (removing data from memory). 1 indicates reading, 0 indicates writing. It might additionally have a clock line or lines for timing, synchronisation, "interrupts," "reset," etc. A microprocessor typically has 10 control lines. Without these essential control signals, the system cannot operate properly.

Control signals are partially unidirectional and partially bidirectional on the Control Bus."Read or write" is an example of a control signal. These signals to memory that we are either reading from or writing to a place that has been specified on the address bus. To regulate and coordinate the system's operation, many more signals are sent. The buses on modern microprocessors like the 80386, 80486 are substantially bigger. Buses that are typically 16 or 32 bits wide, allowing for more memory and faster arithmetic as well as a greater number of instructions.

Micro controllers are organised in the same way, with the exception that because they include memory and other components built into the chip, the buses might all be internal. With the exception of the internal data bus, the three buses in the microprocessor are external to the chip. Buffers, a straightforward electronic link between an external bus and an internal data bus, are used by the chip to connect to external buses when there are any.

20.3 8085A PIN DESCRIPTION

A Figure 20.2 depicts the pin diagram of the 8085A microprocessor. This chip, which has 40 pins, was the final 8-bit general-purpose CPU produced by Intel. The attributes are: 4 Vectored Interrupts, 1 Non Maskable Interrupt, Single + 5V Supply 64K bytes of memory, Serial In/Out Port, Decimal, Binary, and Double Precision Arithmetic Direct Addressing

Capability While the data bus only needs 8 pins and the address bus needs 16, Intel smartly decided to multiplex these two buses so that the data bus shares the lowest 8 pins (A0-A7) of the address bus. Since address and data are never on the bus at the same time, this had no negative effects. The purpose of each pin is described as follows:

Data and Address bus signals

bus signals-AD0-AD7, A8-A15
1) 16 address lines, 2 sets
2) Single-directional most important bits (A8-A15)
It is utilised as both an address bus and a data bus. 3) Bidirectional least significant bits (AD0-AD7)
4) Multiplexed with the bits of a bi-directional data bus

The initial clock cycle of a machine state causes the lower 8 bits of the memory address (or I/0 address) to appear on the multiplexed address/data bus. During the second and third clock cycles, it then switches to being the data bus. In the Hold and Halt modes, 3 was displayed. The eight crucial bits of a memory address or an I/0 address, known as the address bus, are three-stated in hold and halt modes.

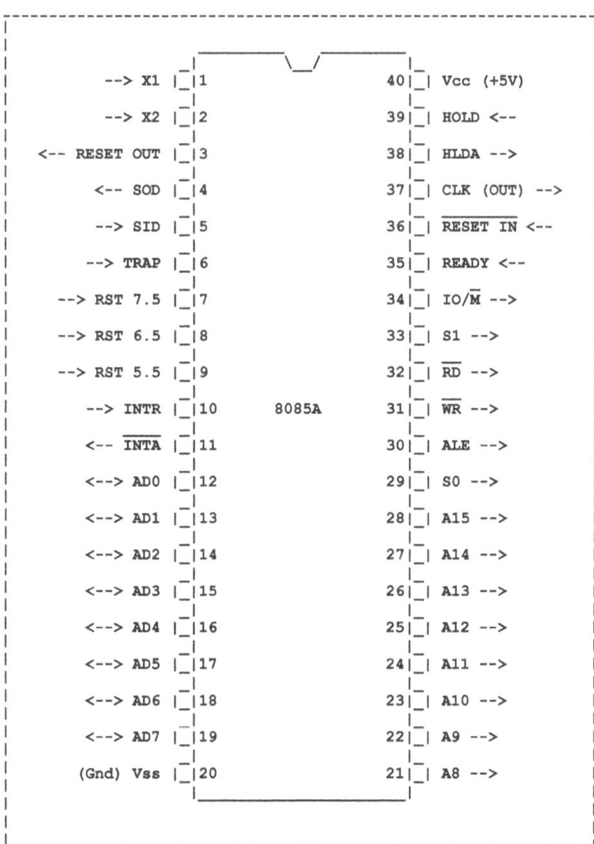

fig. 20.2 Pin diagram of 8085/8085A microprocessor

Toggle Address Latch Enable -ALE (Output)

This happens during a machine state's initial clock cycle and allows the address to be latched into any peripherals' on-chip latches. To ensure setup and hold times for the address information, the falling edge of ALE is configured. Strobe the status information can also be done with ALE. A 3 stated ALE is never used.

Data Bus Status S0, S1 (Output).

Bus cycle's encoded status

S1	S0	
0	0	HALT
0	1	WRITE
1	0	READ
1	1	FETCH

Advanced R/W status S1 is a possibility

RD (State of Output 3)

It also signals that the Data Bus is open for data transfer and that the selected memory or 1/0 device is to be read.

WR (State of Output 3)

WRITE; denotes that the designated memory location or 1/0 position is where the data on the Data Bus is to be written. 3 expressed throughout the Hold and Halt modes. At the WR's trailing edge, data is configured.

A READY (Input)

The memory or peripheral is prepared to send or receive data if Ready is high during a read or write cycle. If Ready is low, the CPU will hold off on finishing the read or write cycle until Ready is high.

The HLDA (Output)

HOLD ACKNOWLEDGE signal tells us that the CPU has received the Hold request and will release the buses in the following clock cycle. Once the hold request has been withdrawn, HLDA drops. Following the low state of HLDA, the CPU uses the buses for one half clock cycle.

INTR (Input) INTERRUPT REQUEST

A general-purpose interrupt is INTR (Input) INTERRUPT REQUEST. Only the next to last clock cycle of the instruction is used to sample it. The Programme Counter (PC) will be prevented from increasing if it is active, and an INTA will be sent. A RESTART or CALL instruction might be introduced during this cycle to jump to the interrupt service routine. Software is used to enable and disable the INTR. Reset and the instant an interrupt is accepted both disable it.

INTA (Output) INTERRUPT ACKNOWLEDGE

After an INTR is acknowledged, INTA (Output) INTERRUPT ACKNOWLEDGE is used in place of (and with the same time as) RD during the Instruction cycle. It can be used to open an interrupt port, such as the 8259 interrupt chip.

(Inputs) RST 5.5 RST 6.,RST 7.5

RESTART INTERRUPTS; These three inputs have the same time as I NTR, with the exception that they automatically insert an internal RESTART.Highest Priority RST 7.5RST 5.5 Lowest Priority RST 6.5

These interruptions are prioritised in the manner previously mentioned. These interruptions are more important than the INTR.

Trap interrupt

A nonmaskable restart interrupt is the TRAP (Input) Trap interrupt. It is acknowledged concurrently with INTR. Neither Interrupt Enable nor any mask have any impact on it. It is the interrupt with the highest priority.

RESET IN (Input) Reset resets the Interrupt Enable and HLDA flipflops as well as the Programme Counter to zero. Except for the instruction register, none of the other flags or registers are impacted. As long as Reset is applied, the CPU will remain in the reset state.

The output RESET OUT (Output) indicates that CPlJ is reset. can be used to RESET a system. The signal and processor clock are in sync.connections for the X1, X2, or (Input) Crystal or R/C networks to set the internal clock generator In addition to being a crystal, X1 can also be an external clock input. The internal operating frequency is calculated by dividing the input frequency by two.

(Output) CLK Clock When a crystal or R/C network is utilised as an input to the CPU, the output is used as the system clock. Two times as long as the X1, X2 input period is CLK's period.

the output, IO/M

When in Hold and Halt modes, IO/M specifies whether the Read/Write is to memory or l/O Tristated.

Serial input data line SID (Input) Every time a RIM command is carried out, the information on this line is placed into accumulator bit 7.

Serial output data line, or SOD. Depending on what the SIM instruction specifies, the output SOD is set or reset.

Supply voltage: +5.Vcc

Ground Reference for. Vss

signals that are sent and acknowledged from outside the system commenced signals CPU reset using Reset In Hold to halt CPU activity to sync with slower devices, the CPU enters the ready state, acknowledgment signal.

Once the CPU is at rest, reset out to high. HLDA - recognises the hold signal.

20.4   WORKING OF 8085A

Fig. 20.3 depicts the entire 8085A 8-bit parallel central processor. It needs a single supply of +5 volts. With a faster system speed and a fundamental clock speed of 3 MHz, it performs better than the 8080 than is currently in use. Additionally, it is made to fit into systems with a minimum of three ICs: a CPU, a RAM/IO chip, and a ROM or PROM/IO chip.The Data Bus is multiplexed on the 8085A. The lower 8bit Address/Data Bus and the higher 8bit Address Bus divide the address into two parts. The first cycle is when the address is distributed. The Address Latch Enable (ALE) latches the lower 8 bits into the peripherals. Memory is used on the Data Bus for the remainder of the machine cycle.

Bus control signals for the 8085A include RD, WR, and lO/ Memory. There is also a signal called the Interrupt Acknowledge (INTA) signal. Every interruption—Hold, Ready, and Ready—is timed in unison. A straightforward serial interface is also provided by the 8085A with serial input data (SID) and serial output data (SOD) lines. The 8085A also includes three restart interrupts that are maskable and one trap interrupt that is not maskable. Bus control is accomplished via the 8085A's RD, WR, and IO/M signals.

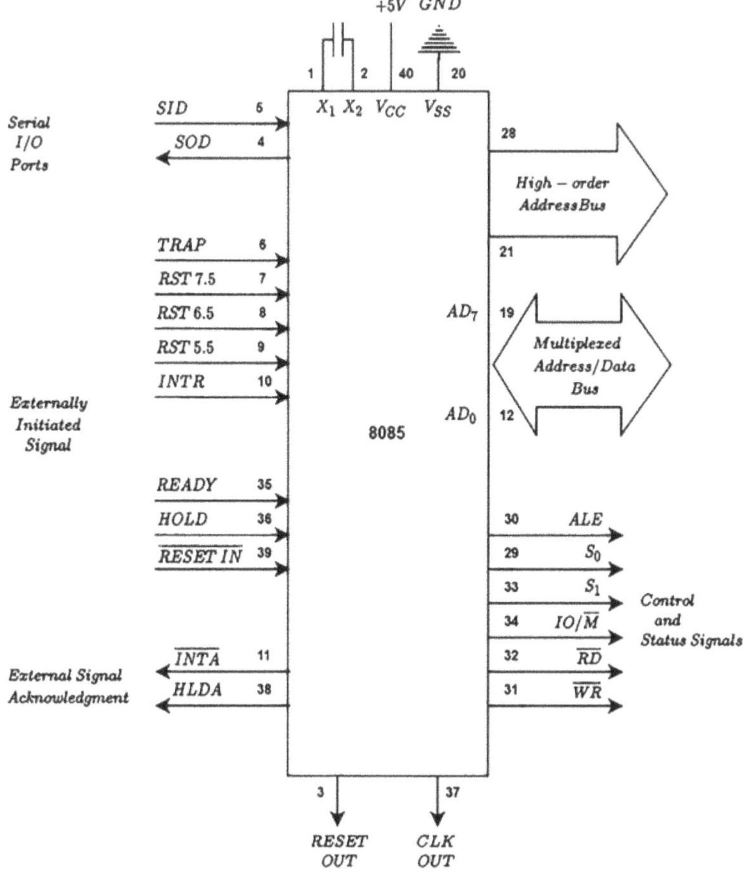

fig. 20.3 Functional diagram of 8085/8085A microprocessor

Condition Reports

The 8085A immediately provides status information. As a status strobe, ALE is used. The status, which is partially encoded, gives the user advance timing information about the type of bus transfer being performed. Also directly given is the IO/M cycle status signal. Decoded The status information carried by S1 is as follows:

Stop, Write, Read, and Fetch

In all bus transfers, S1 is equivalent to R/W. In the 8085A, the data is multiplexed with the 8 LSB of the address rather than the status. To input the lower half of the address into the memory or peripheral address latch, the ALE line is employed as a strobe. Additionally, this frees up extra pins for increased interrupt capability.

l/O interrupts and serial

Five interrupt inputs are available on the 8085A: INTR, RST5.5, RST6.5, RST7.5, and TRAP. The 8080 INT and INTR both perform the same tasks. The 5.5, 6.5, and 7.5 RESTART inputs each include a configurable mask. While TRAP is a RESTART interrupt, it cannot be muted. The three RESTART interrupts, which save the programme counter in the stack and branch to the RESTART address, induce internal execution of RST if interrupts are enabled but the interrupt mask is not set. Independent of the state of the interrupt enable or masks, the non-maskable TRAP triggers the internal execution of a RST.

If there are many pending interrupts, the interrupt with the highest priority is recognised. The interrupts are ordered as follows: RST 7.5, RST 6.5, RST 5.5, TRAP highest priority, INTR lowest importance The priority of a routine that was initiated by an interrupt with a higher priority is not taken into account by this priority scheme. If the interrupts were enabled again before the RST 7.5 function ended, RST 5.5 may interrupt it. For catastrophic errors like a power outage or a bus issue, the TRAP interrupt is helpful. The TRAP input has the highest priority but is nevertheless recognised like any other interrupt. No flag or mask has any effect on it. Both edges and levels are responsive to the TRAP input.

Timing for a Basic System

The Data Bus of the 8085A is multiplexed. To sample the lower 8 bits of address on the Data Bus, ALE is used as a strobe. Figure 2 depicts a memory read, l/ O write, and instruction fetch cycle. The l/O port address is copied on both the upper and bottom halves of the address during the l/O write and read cycle, it should be noted. Like the 8080, the 8085A can be utilised with sluggish memory by extending the read and write pulse lengths via the READY line. When the CPU is finished utilising a bus, hold causes the Address and Data Buses to float, causing the CPU to release the bus.

System Interface 8085A series.

Memory parts that are directly compatible with the 8085A CPU are part of the System Interface 8085A series. The following characteristics, for instance, will be present in a system made up of the three chips 8085A, 8156, and 8355:

2K bytes of ROM, 256 bytes of RAM, a timer/counter, four 8-bit I/O ports, one 6-bit I/O port, four interrupt levels, and serial in/out portsA useful method of l/O addressing is provided by memory mapped I/O in addition to normal l/O. With this method, a portion of the memory address space is designated for I/O addresses, allowing for the manipulation of memory addresses for I/O. The conventional memory, which lacks a multiplexed address/data channel, can also communicate with the 8085A CPU. address/data bus.

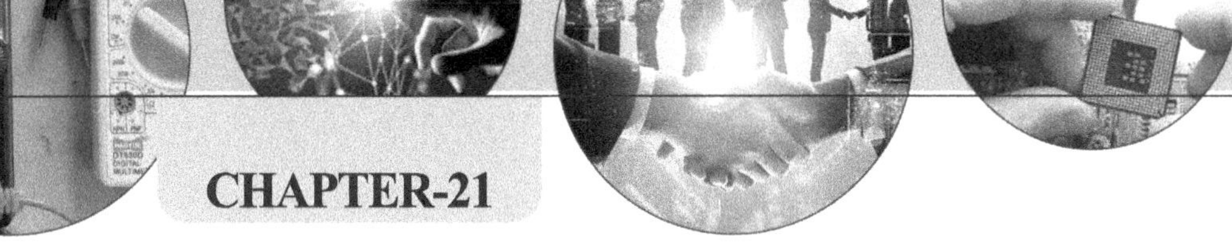

CHAPTER-21

Programming model of 8085 microprocessor

2.1.1 INTRODUCTION

A computer can only carry out the instructions provided by the programmer in the form of a programme. A programme is simply a collection of straightforward instructions that directs the computer to carry out a certain task. The computer can execute the instructions very quickly and consistently and without making a mistake once the programme has been created and debugged (you hardly ever get it correctly the first time).

And this is where a computer really shines. Even though the programme only only a few straightforward instructions, the end product can nevertheless be very remarkable, largely because of how quickly the computer can execute those instructions. Even though the program's phases are all quite straightforward, the sequence of instructions, which is carried out at a rate of millions of steps per second, can appear to be highly complicated when viewed as a whole. The secret is to break it down into a succession of incredibly basic commands rather than viewing it as a whole.

Typically, a single integrated circuit (IC) serves as the microprocessor. The orders or instructions that can be executed by most microprocessors—also known as very small computers—are referred to here simply as "micros." They differ mostly in the names given to each command. In a typical micro, there are commands to move data about, add, subtract, multiply, and divide, as well as to bring data into the micro from the outside world and transfer information from the micro out into the world.

Three fundamental components make up a standard microprocessor. Input/Output (I/O), Memory, and the Programme Counter (PC) are these components. Which command will be executed next is kept track of by the programme counter. The commands to be carried out are stored in the Memory. The transfer of data to and from the outside world (apart from the micro's physical enclosure) is handled by the input/output system. The 40 pin package, chip, or IC that houses the micro we'll be using.

A sequence or collection of straightforward commands or instructions makes up a programme. The issue of filling a cup with liquid from a tap could serve as a practical example programme.

Step 1: Turn on the liquid.

Step 2: Put the cup under the tap.

Step 3: Look at the cup.

Step 4: Is it full?

5. If not, proceed to step 3

Step 6: Take out the cup from beneath the tap.

Step 7: Turn off the liquid. A programmer has two options for programming a Processor, assembly language and machine language.

According to Fig. 21.1, the 8085 programming model has six registers, an accumulator, and a flag register. It also contains two 16-bit registers, the programme counter and the stack pointer. A quick summary of them is provided below.

ACCUMULATOR A (8)		FLAG REGISTER	
B (8)		C	(8)
D (8)		E	(8)
H (8)		L	(8)
Stack Pointer (SP)			(16)
Program Counter (PC)			(16)

8 Lines Bidirectional   16 Lines unidirectional

Data Bus   Address Bus

fig. 21.1 Registers

21.1.1 Registers

Eight-bit data can be stored in the 8085's six general-purpose registers, which are labelled B, C, D, E, H, and L as indicated in the image. For some 16-bit operations, they can be coupled as register pairs: BC, DE, and HL. By employing data copy instructions, the programmer can use these registers to store or copy data into the registers.

21.1.2 Accumulator

An 8-bit register called the accumulator is a component of the ALU, or arithmetic/logic unit. In addition to storing and retrieving 8-bit data, this register is utilised to carry out arithmetic and logical operations. An operation's result is kept in the accumulator. The register A is another name for the accumulator.

21.1.3 Flags

The accumulator and other registers' data conditions determine whether the five flip-flops in the ALU are set or reset following an operation. These flags are designated as Zero (Z), Carry (CY), Sign (S), Parity (P), and Auxiliary Carry (AC); Fig. 21.2 shows their bit positions in the flag register. The three flags Zero, Carry, and Sign are the most often employed. These flags are tested by the microprocessor to determine data conditions.

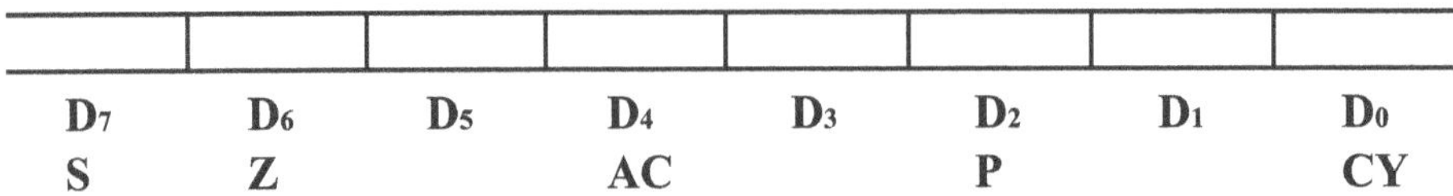

D_7	D_6	D_5	D_4	D_3	D_2	D_1	D_0
S	Z		AC		P		CY

fig. 21.2 Flags

The carry flag (CY), which the flip-flop employs to signal a carry, is set to one when, for instance, the sum of the two values added in the accumulator exceeds eight bits. The flip-flop designated as the Zero(Z) flag is set to one whenever an arithmetic operation yields a result of zero. An 8-bit register known as the flag register is seen next to the accumulator in the first Figure. Five of the eight bit places are not used as registers; instead, they are used to store the outputs of the five flip-flops.

The flags are kept in the 8-bit register so that the programmer can access the register using an instruction to check the flags (data conditions). The microprocessor's decision-making process heavily relies on these flags. Through the programme instructions, the flags' conditions (set or reset) are tested. For instance, when the CY flag is set, the instruction JC (Jump on Carry) is used to alter the program's sequence. Writing assembly language programmes requires a solid understanding of flag.

21.1.4   A Programme Counter, or PC

This 16-bit register manages the order in which instructions are carried out. A memory pointer exists in this register. This register has 16 bits because memory locations have 16-bit addresses. This register is used by the P to order how the instructions are executed. The programme counter's job is to indicate the memory location where the subsequent byte should be fetched. The programme counter is shifted to the following memory address each time a byte of machine code is fetched, increasing the counter by one.

21.1.5 A stack pointer, or SP

A 16-bit register is also utilised as a memory pointer by the stack pointer. It identifies the stack, a memory region in R/W memory. By putting a 16-bit address into the stack pointer, the start of the stack is established. In later courses, this programming model will be used to analyse how these registers are impacted by the execution of instructions.

21.2   A DECIMAL, BINARY & HEX SYSTEMS

The systems used by the Decimal, Binary, and Hex microprocessors differ. Our computers are known as binary computers since it is called binary. The same guidelines apply to all coding schemes. Base 10 is used for decimal, Base 2 for binary, and Base 16 for hexadecimal. The number of numbers that can fit in each digit place is referred to as a system's base. A single digit number in decimal is one from 0 to 9. A single digit integer in binary is either 0 or 1. A single digit number in hex is represented by 0 through 9, A, B, C, D, E, and F.

21.3   INSTRUCTIONS FOR MACHINE LANGUAGE

A central processing unit (CPU) or microprocessor contains a variety of Registers. They differ from CPU to CPU, but they are all comprised of a register known as the Accumulator. Some people also refer to it as the A registry. The accumulator will be used in the discussion that follows. It is a form of temporary memory that is 8 bits wide, or a byte, like the majority of locations where data can be stored inside the CPU.

The CPU we'll be using has 5 different sorts of instructions, each with many variations, totaling over 100 distinct instructions. They are as follows:

a) Arithmetic operation b) Logical operation c) Data transfer operations d) branching operations

21.3.1 Arithmetic Operations

The usual arithmetic operations include addition, subtraction, division, multiplication, incrementing, and decrementing, albeit most early CPUs did not support division or multiplication. The programme can be informed of the results of an instruction using two flags when using arithmetic. The Carry (C) flag is the first. The Z flag stands in for zero. In the subsequent addition example, the C flag will be described. If the Z flag is set, the operation's outcome left the accumulator with a value of 0, which is indicated by the instruction. Later in the class, we'll see how the Z flag is used.

21.3.1.1 Addition

This is simple; all that is required is to add two integers to obtain the answer. There is one more thing, though. A portion of the result might be lost if the sum was too large to fit inside the accumulator. A defence exists against this. Consider the binary numbers 11111111 and 11111111, both of which are 255. The greatest numbers that can be stored in an 8-bit register or memory address are these. Since I also provided you with their decimal values, you can add these two to get 510. 510 has a binary value of 111111110 (9 bits).

The accumulator, which is a byte, is only 8 bits wide. How can you cram a 9-bit number into an 8-bit container? You cannot, and the situation is referred to as an OVERFLOW condition. We use the CARRY (C) flag for this. The CARRY (C) flag will hold the ninth bit if the sum of the addition's results is more than 8 bits. In this instance, the C flag would be a 1, or set, and the accumulator would contain the value 11111110 (254) in it. Because it is the ninth bit, this 1 has the value of 256.

21.3.1.2 Subtraction

It involves the representation of the compliments of 1 and 2. But I will say that if you subtract two values, the C flag will be a 1 if there is an underflow, otherwise it will be a 0. A under flow occurs when a greater number is subtracted from a smaller one.

A number's 1's complement is created by deducting 1 from each bit. The 1's complement of 1100, for instance, is 0011. You can easily obtain a binary number's 1's complement by flipping each bit, or by turning a 1 to a 0 and a 0 to a 1. A binary number's 2's complement is equal to its 1's complement + 1. For instance, 0011+1=0100 is the 2's complement of the number 110.

21.3.1.3 Dividing and multiplying

The multiply and division instructions are not present in the microprocessor we'll be utilising, the 8085. Processor will execute addition for multiplication and division will be subtracted.

21.3.1.4 Decrement and Increment

The group of arithmetic instructions also contains two more instructions. Both increment and decrement apply. A programme can use these instructions to count occurrences or loops.

When an increment is carried out, the value is increased by 1 each time. By one, a decrement reduces the value. These can be used in conjunction with conditional jumps to repeatedly loop a block of code.

21.3.2　The Logical Operation

There are more mathematical instructions in Processor referred to as logical instructions. There are eight of them: OR, AND, XOR, ROTATE, COMPLEMENT, and CLEAR. These instructions typically don't care about the value of the data they deal with, but rather the value or condition of each individual bit.

21.3.2.1 OR OR By using the binary integers 1010 and 0110, the OR function can be demonstrated. It makes no difference whether you start on the right or left end when O Ring two numbers. Let's move left and begin. There is a 1 in the first number and a 0 in the second number in the first bit position. This would yield a score of 1. The following bit has two numbers, the first of which is 0 and the second of which is 1. It would result in 1. The next bit has a 1 in both the first and second numbers. A 1 would result as a consequence. The final bit has a 0 in both the first and second numbers, yielding a 0. So, 1110 would be the response. According to the rule that produces this outcome, a 1 in either number produces a 1 with an OR, or to put it another way, any 1 in produces a 1 out.

21.3.2.2　AND Different rules apply when AND'ing. For each equivalent bit position, the rule is that a 0 in either number will result in a 0 in the output. The answer would be 0010 if you used the same two numbers, 1010 and 0110. It is evident that every bit position other than the third has a zero in one of the numbers. The result of a 1 AND a 1 is a 1 is another way to describe an AND.

21.3.2.3　Exclusive OR XOR with one exception, XORing is akin to ORing. An inclusive OR is another name for an OR. Thus, a 1 in any or both of the numbers will equal a 1. A result of 1 will be produced by an exclusive OR if either number contains a 1, but not both. A apparently insignificant but important distinction. The outcome would be 1100 if the same two numbers were used. The first or second number in the first two bits has a 1, but not both. Because both numbers have a 1 in the third bit, the outcome is 0. The result of the fourth is 0, as there are absolutely no 1s. Despite the fact that the OR and XOR yield distinct results, the difference may appear to be slight. The main function of an XOR is to compare two numbers. If they are identical, the response will be all 0s; if they differ, the answer will be all 1.

21.3.2.4　Compliment A number that is complimented takes on all the 1s and 0s in the opposite state. Consider the digits 1111. Positive outcomes come in 0000. The simplest and most understandable operator is this one. You'll discover later that it has a variety of functions, all of which are essential.

21.3.2.5　Rotate Bits of a byte are rotated using these instructions. Each instruction rotates one bit in either the left or right direction. An illustration would be if the accumulator contained the number 11000011.

The outcome of rotating to the left is 100001111. As you can see, bit 7 has now been relocated to bit 0, and every other bit has shifted one bit position to the left.

21.3.2.6 Clear -This command zeroes out or clears the accumulator. Moving a 0 into the accumulator is equivalent to doing this. Additionally, the Z flag is set and the C flag is cleared.

21.3.3 The Operation of Branching

This set of instructions modifies how a programme executes, either conditionally or always.

21.3.3.1 Jump-In the programming, conditional jumps play a significant role in the decision-making process. These instructions check whether specific conditions are true (such a zero or a carry flag), and when they are, they change the program's flow. Additionally, there is a command called unconditional jump in the instruction set.

21.3.3.2 Call, Return, and Restart-These commands alter a program's flow by either calling a subroutine or returning from one. Condition flags can also be tested using the conditional Call and Return instructions. be used by the primary programme whenever it need it, serving as a means of generating custom instructions.

21.3.4 Moving with the data transfer operation

These instructions do exactly what you would think. They move data around between the various registers and memory.

Exchanging

A version of the move instruction is exchanging. In this instance, two locations exchange data.

Operations Controlling Machines

These commands command the machine to do things like Halt, Interrupt, or do nothing. Four functions can be used to sum up the microprocessor activities involved in data manipulation:1. duplicating data; 2. carrying out calculations3. doing out logical operations.4. examining a specific circumstance and alerting the program sequence.

The following list highlights some key elements of the instruction manual:

1. When data is transferred, the source's contents are not lost; instead, only the destination's contents are altered. Flags are not impacted by the data copy instructions.

2. The contents of the accumulator are subjected to arithmetic and logical operations, and the outcomes are recorded (with some expectations) in the accumulator. According on the findings, the flags are changed.

3. Incrementing and decrementing can be done with any register, even the memory.

4. A program sequence can be modified by testing for a specific data condition or by conditionally changing the programme.

CHAPTER-22

8085/8085A µP
Assembler Language Programming

22.1 PROGRAMMING INSTRUCTIONS

Your text-based source file is changed into a machine language file by an assembler. The built-in text editor is used to construct the source file. The source file contains the machine language instructions that the assembler transforms from the source code into a string of integers.

TASK	Op code	Operand	Binary Code	Hex Code
Copy the contents of the accumulator in the register C.	MOV	C,A	01001111	4FH
Add the contents of register B to the contents of the accumulator	MOV	B	1000 0000	80H
Invert (compliment) each bit in the accumulator	CMA		0010 1111	2FH

After the numbers are memorised, the memory Three different tasks are carried out by these one-byte instructions. Both operand registers are given in the first instruction. The operand B is provided and the accumulator is presumed in the second instruction. The accumulator is also thought to be the implicit operand in the third instruction. These instructions are kept in memory in 8-bit binary code, and each one needs a specific memory location.

22.1.1: Instructions in Two Bytes

In a two-byte instruction, the operand is specified in the second byte, while the first byte contains the operation code. A data byte comes right after the opcode and is known as the source operand.

For instance:Count on the data byte being 32H. The instruction is written in assembly language asTwo memory locations would be needed to store the instruction in memory.

TASK	Op code	Operand	Binary Code	Hex Code	
Load an 8-bit data byte in the accumulator.	MVI	A₂ Data	0011 1110	3E	First Byte
			DATA	DATA	Second byte

22.1.2 Instructions in Three-Byte

The first byte, which provides the opcode, and the second and third bytes, which specify the 16-bit address. Keep in mind that the second byte is the low-order address and the third byte is the high-order address. data byte plus data byte plus opcode

For instance, storing this instruction in memory would take three memory addresses. Opcode + Data + Data are the three bytes that make up an instruction.

TASK	OPCODE	OPERAND	BINARY CODE	HEX CODE	
Transfer the program sequence to the memory location 2085H.	JMP	2085H	1100 0011	C3	First byte
			1000 0101	85	Second Byte
			0010 0000	20	Third Byte

22.2 INSTRUCTION SET 8085

22.2.1 Directions for data transfer

MOV Copy from the source to the MVI destination Move right away. an 8-bit LDALDAX, or load accumulator Load register pair immediately LXI Load accumulator indirectly

LHLD Directly load the H and L registers.

Direct STA Store accumulator

Direct SHLD Store H and L registers, STAX Store accumulator

H and L are exchanged for D and E in XCHG.

SPHL Copy the stack pointer's H and L registers.

XTHL Swap H and L at the top of the stack.

PUSH Stack POP: Push register pair on to Stack pop to register pair OUT Data output from the accumulator should go to a port with the IN 8-bit address. Fill the accumulator with data from a port with an 8-bit address.

22.2.2 Arithmetic Directives

ADD   Add register or memory to accumulator
ADC   Add register and carry to the accumulator
ADI   Add immediate to accumulator
ACI   With carry, add immediate to the accumulator
DAD   To the H and L registers, add a register pair
SUB   Register or memory subtraction from accumulator
SBB   Subtract the source and borrow from the accumulator,
SUI   Subtract immediate from accumulator
SBI   With borrow, subtract immediate from accumulator
INR   Register or memory increment by 1
INX   Increment register pair by 1
DCR   Decreasing a memory or register by 1
DCX   Decrement register pair by 1
DAA   Decimal adjust accumulator

22.2.3 The Control Directives

NOP   No operation
HLT   Halt
DI   Disable interrupts
EI   Enable interrupts
RIM   Read interrupt mask
SIM   Set interrupt mask

22.2.4 Directions For Branching

JMP   Jump unconditionally
JC   Jump on carry
JNC   Jump on no carry
JP   Jump on positive
JM   Jump on minus
JZ   Jump on zero
JNZ   Jump on no zero
JPE   Jump on parity even
JPO   Jump on parity odd
CALL   Call unconditionally
CC   Call on carry
CNC   Call on no carry
CP   Call on positive
CM   Call on minus
CZ   Call on zero

CNZ   Call on no zero
CPE   Call on parity even
CPO   Call on parity odd
RET   Return unconditionally
RC   Return on carry
RNC   Return on no carry
RP   Return on positive
RM   Return on minus
RZ   Return on zero
RNZ   Return on no zero
RPE   Return on parity even
RPO   Return on parity odd
PCHL   Load program counter with HL contents
RST   Restart

22.2.5   Instructions for logical operation

CMP   Comparing an accumulator to a register or memory
CPI   Compared to the accumulator, the immediate
ANA   logical AND memory or a register with an accumulator
ANI   logical AND memory or a register with an accumulator
XRA   Exclusive OR memory or a register with an accumulator
XRI   Exclusive OR instantaneous with accumulator
ORA   register or memory with an accumulator or logical OR
ORI   logical OR instantaneous with accumulator
RLC   Turn the accumulator left
RRC   Turn the accumulator right
RAL   Rotate the accumulator to the left through carry
RAR   Turn the accumulator through carry on the right.
CMA   accumulator of complements
CMC   Carry in complement
STC   Set carries

22.3   INSTRUCTION DETAILS FOR 8085

22.3.1   Instructions for data transfer

Opcode   **Operand**   **Description**

To copy source to destination

MOV   Rd, Rs   copy from the source to the target The source
        M, Rs   register's content are not changed by this instruction;
        Rd, M   they are simply copied into the destination register. The information in the HL registers determines where an operand is located in memory if it is one of the operands. For instance, MOV B, C or MOV B, M

Move immediately 8-bit

MVI	Rd, data	The final register or memory receives the 8-bit data and M, data stores it there. If the operand is a memory location, the HL registers' data will tell you where it is if it is a memory location.For instance, MVI M, 57H or MVI B, 57H

Load for accumulator

LDA	16-bit address	A 16-bit address in the operand specifies the memory location, and the contents of that location are copied to the accumulator. The source's material is not changed in any way. Consider LDA 2034H.

Load for accumulator indirect

LDAX	B/D Reg. pair	The selected register pair's contents identify a memory location. The information in that memory location is copied by this instruction into the accumulator. There is no change to the information in the memory location or the register pair. Consider LDAX B.

Immediate load register pair

LXI	Reg. pair, 16-bit	The instruction fills the register pair specified in the argument with 16-bit data.For instance, LXI H, XYZ or LXI H, 2034H Directly load H and L registers
LHLD	16-bit address.	The instruction copies the data from the memory location indicated by the 16-bit address into register L before moving on to register H to copy the data from the following memory location. The information in the source memory locations is not changed. Consider LHLD 2040H.direct accumulator storage
STA	16-bit address	The operand-specified memory location receives a copy of the accumulator's data. The second byte of this 3-byte instruction indicates the low-order address, and the third byte specifies the high-order address. STA 4350H, as an illustration.

Accumulator store indirect

STAX	Reg. pair	The contents of the operand (register pair) are used to specify where in memory to copy the contents of the accumulator. The accumulator's contents are not changed. Consider STAX B.

Direct store H and L registers

SHLD	16-bit address	By increasing the operand, the contents of register H are put in the succeeding memory location, and the contents of register L are stored in the memory location indicated by the operand's 16-bit address. The content of Registers HL is not altered. this 3-byte instruction defines the low-order address in the second byte and the high-order address in the third byte. as in SHLD 2470H

H and L are swapped for D and E.

XCHG   none   Specifically, the contents of registers H and L are swapped for those of registers D and E, respectively.as in XCHG

Copy the stack pointer's H and L registers.

SPHL   none   The H register's contents supply the high-order address, while the L register's contents offer the low-order address, and the instruction loads the contents of both registers into the stack pointer register. The information in the H and L registries is not changed.Instance: SPHL

Swap H and L with the top of the stack

XTHL   none   The stack location indicated by the data in the stack pointer register is exchanged for the contents of the L register. The H register's contents are switched with those of the ubsequent stack location (SP+1), while the stack pointer register's contents are left alone.Instance: XTHL

Stack a register pair by pushing

PUSH   Reg. pair The following steps are used to copy the data from the operand-designated register pair onto the stack. The high order register's contents (B, D, H, and A) are copied into the stack pointer register once it has been decremented. The low-order register's contents (C, E, L, and flags) are duplicated there once the stack pointer register is decremented once more. For instance, PUSH B or PUSH A

To register a pair, lift the stack off.

POP   Reg. pair Remove stack and register a pair.The low-order register (C, E, L, status flags) of the operand receives a copy of the data from the memory location the stack pointer register specifies. It is the stack pointer. The value of that memory address is raised by 1, and its contents are then copied to the operand's high-order register (B, D, H, A). Once more, a one-digit increase is added to the stack pointer register. Example: A POP H or a POP

Data output from the accumulator

OUT   a port with The operand-specified I/O port receives a copy of the accumulator's 8-bit address data. as in OUT F8H Input data to accumulator from a port with 8-bit address IN a port with 8-bit address The accumulator is filled with the contents of the input operand port that the operand specifies being read from. Like this: IN 8CH

22.3.2 Arithmetic directives

Opcode Operand Description

ADD   R   The contents of the accumulator are added to the M contents of the operand (register or memory), M, and the result is saved in the accumulator. If the operand is a memory location, the HL registers' data will tell you where it is if it is a memory location. All flags are changed to reflect the addition's outcome. For instance, ADD M or ADD B

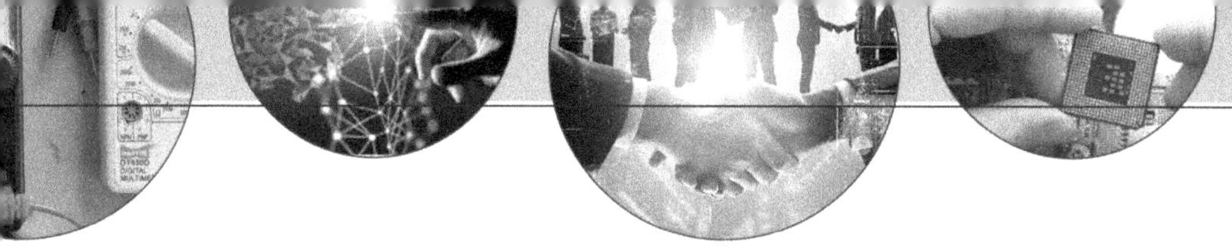

Add register and carry to the accumulator.

ADC   R   The contents of the accumulator are added to the M contents of the operand (register or memory) M and the Carry flag, and the resulting value is stored in the accumulator. If the operand is a memory location, the HL registers' data will tell you where it is if it is a memory location. The result of the addition is reflected in all flags. For instance, ADC B or ADC

Add the instant to the accumulator.

ADI   8-bit data   The accumulator's contents are combined with the 8-bitdata (operand), and the resulting value is then placed there. The result of the addition is reflected in all flags. as in ADI 45H

With carry, add immediate to the accumulator.

ACI   8-bit data   The accumulator's contents are added to the 8-bit data (operand) and the Carry flag, and the resulting value is then stored there. The result of the addition is reflected in all flags. Consider ACI 45H.

Add register pair to H and L registers

DAD   Reg. pair   The HL register's contents are combined with the 16-bit contents of the chosen register pair, and the result is then placed there. The source register pair's contents are not changed. The CY flag is set if the result is more than 16 bits big. Other flags are not impacted. Instance: DAD H

Subtract register or memory from accumulator

SUB   R   The contents of the accumulator are subtracted from the M contents of the operand (register or memory), and the result is saved in the accumulator. If the operand is a memory location, the HL registers' data will tell you where it is if it is a memory location. To reflect the results of the subtraction, all flags are changed. SUB B or SUB M, for instance

Subtract source and borrow from accumulator

SBB   R   The Borrow flag and the contents of the operand (register or M memory) are deducted from the accumulator's contents, and the result is added to the accumulator. If the operand is a memory location, the HL registers' data will tell you where it is if it is a memory location. To reflect the results of the subtraction, all flags are changed. SBB B or SBB M, for instance Subtract immediate from accumulator

SUI   8-bit data   The 8-bit data (operand) is subtracted from the accumulator's contents, and the result is then saved there. To reflect the results of the subtraction, all flags are changed. as in SUI 45H

Subtract immediate from accumulator with borrow

SBI   8-bit data   The Borrow flag and the 8-bit data (operand) are subtracted from the contents of the accumulator, and the result is saved there. Every flag is changed to reflect the outcome of the subtraction. as in SBI 45H

Increment register or memory by 1

INR   R   The chosen register's or memory's contents are increased by M1, and the outcome is saved there as well. If the operand is a memory location, the HL registers' data will tell you where it is if it is a memory location. For instance, INR B or INR M

Opcode   Operand Description

Increment register pair by 1

INX   R   The chosen register pair's contents are increased by 1 and the result is saved there as well. Instance: INX H

Decrement register or memory by 1

DCR   R   The result is put in the same location after the contents of the M chosen register or memory M are reduced by 1. If the operand is a memory location, the HL registers' data will tell you where it is if it is a memory location. DCR B or DCR M, as examples

Decrement register pair by 1

DCX   R   Add one DCX R to each register pair. Decreasing by 1 the contents of the chosen register pair and saving the outcome in the same place

Decimal adjust accumulator

DAA   none   Two 4-bit binary coded decimal (BCD) digits are substituted for a binary value as the contents of the accumulator. This is the only instruction that converts from binary to BCD using the auxiliary flag; the conversion process is detailed below. The S, Z, AC, P, and CY flags are changed to reflect the operation's outcomes. The instruction adds 6 to the low-order four bits if the value of the accumulator's low-order 4-bits is larger than 9 or if the AC flag is set. The instruction adds 6 to the high-order 4-bits if the accumulator's high-order 4-bit value is more than 9 or if the Carry flag is set. Example: DAA

22.3.3 Branching Instructions

Jump unconditionally

JMP   16-bit address   The 16-bit address provided in the operand is used to specify the location in memory where the programme sequence should be moved. For instance, JMP 2034H or JMP XYZ

Jump conditionally

Operand: 16-bit address   Based on the stated flag of the PSW, the programme sequence is transferred to the memory location indicated by the 16-bit address provided in the operand as explained below.For instance, JZ 2034H or JZ XYZ

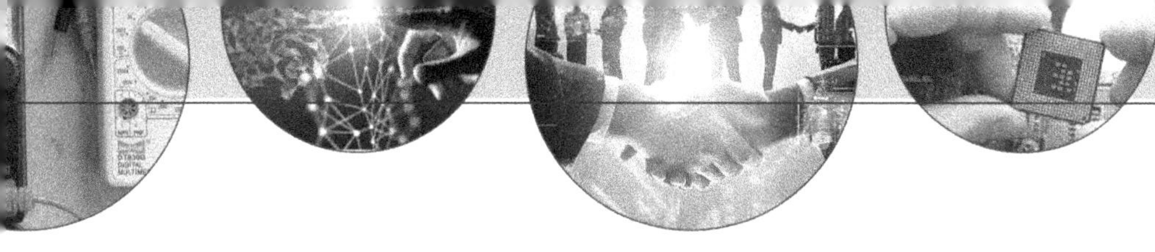

Opcode	Description	Flag Status
JC	To Jump on Carry	CY = 1
JNC	To Jump on no Carry	CY = 0
JP	To Jump on positive	S = 0
JM J	To Jump on minus	S = 1
JZ	To Jump on zero	Z = 1
JNZ	To Jump on no zero	Z = 0
JPE	To Jump on parity even	P = 1
JPO	To Jump on parity odd	P = 0

Unconditional subroutine call

CALL 16-bit address   The 16-bit address provided in the operand is used to specify where in memory to move the programme sequence. The programme counter's contents, or the address of the following instruction after CALL, are pushed into the stack prior to the transfer. For instance, dial 2034H or XYZ.

Call conditionally

Operand: 16-bit address   Based on the stated flag of the PSW, the programme sequence is transferred to the memory location indicated by the 16-bit address provided in the operand as explained below. The programme counter's contents, or the address of the following instruction after the call, are placed into the stack prior to the transfer. For instance, CZ 2034H or CZ XYZ

Opcode	Description	Flag Status
CC	To Call on Carry CY = 1	
CNC	To Call on no Carry	CY=0
CP	To Call on positive	S = 0
CM	To Call on minus	S = 1
CZ	To Call on zero	Z = 1
CNZ	To Call on No Z = 0.	Z = 0
CPE	To Call on Parity even if	P = 1
CPO	To Call on Parity odd if	P = 0

Return from subroutine unconditionally

RET   none   The calling programme receives the programme sequence that was passed from the subroutine. At the new address, programme execution starts after copying the top two bytes of the stack into the programme counter. Instance: RET

Return from subroutine conditionally

Operand: none   Using the PSW's specified flag and the following explanation, the programme sequence is passed from the subroutine to the caller programme. Execution of the programme starts at the new address once the two bytes from the stack's top are copied into the programme

Counter.Examples: RZ

Opcode	Description	Flag Status
RC	To Return on Carry	CY = 1
RNC	To Return for carry nill	CY = 0
RP	To Return on positive	S = 0
RM	To Return on minus	S = 1
RZ	To Return on zero	Z = 1
RNZ	To Return on no zero	Z = 0
RPE	To Return on parity even	P = 1
RPO	To Return on parity odd	P = 0

Load program counter with HL contents

PCHL   none   The programme counter receives copies of the contents of registers H and L. The high-order byte is represented by the contents of H, and the low-order byte is represented by the contents of L. Instance: PCHL

Restart

RST   0-7   Depending on the number, the RST instruction corresponds to one of eight 1-byte call instructions in memory. Typically, interrupts are used in conjunction with the instructions, which are added using external hardware. To move programme execution to one of the eight places, these can be utilised as software instructions in a programme. the following addresses:

Instruction	Restart Address
RST 0	0000H
RST 1	0008H
RST 2	0010H
RST 3	0018H
RST 4	0020H
RST 5	0028H
RST 6	0030H
RST 7	0038H

There are four extra interrupts on the 8085, and they internally generate RST instructions without the need of any additional hardware. These guidelines and their respective Restart URLs are:

Interrupt	Restart Address
TRAP	0024H
RST 5.5	002CH
RST 6.5	0034H
RST 7.5	003CH

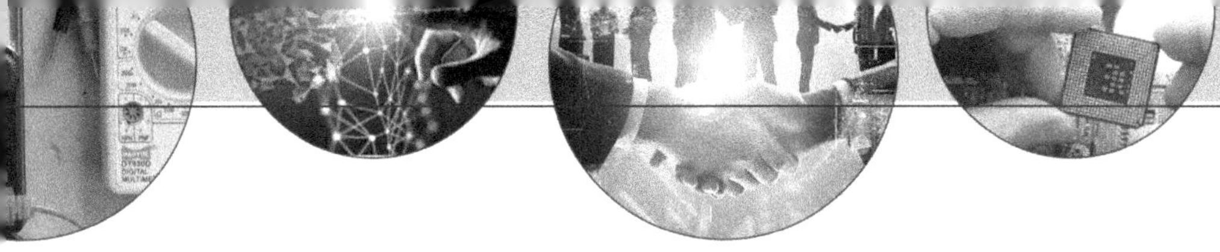

22.3.4 Logical Instructions

Opcode Operand Description

Compare register or memory with accumulator

CMP   R   The accumulator's contents are compared to the   M operand's (register or memory) contents. There is no loss to either content. By configuring the PSW flags in the following manner, the comparison's outcome is displayed: If (A) (reg/mem), the carry flag is set. If (A) = (reg/mem), the zero flag is set. If (A) > (reg/mem), both the carry and zero flags are reset. Example: CMP M versus CMP B

Compare immediate with accumulator

CPI   8-bit data   A comparison is made between the second byte's (8-bit) data and the information in the accumulator. The values under comparison don't change. The comparison's outcome is displayed by setting the PSW flags as follows: the carry flag is set if (A) data If (A) = data, the flag is set to zero. If (A) > The carry and zero flags are reset in the data. CPI 89H, for instance

Logical AND register or memory with accumulator

ANA   R   The result is added to the accumulator after the M contents of the accumulator and operand (a register or memory) have been logically ANDed. If the operand is a memory location, the HL registers' contents serve as the operand's address. S, Z, and P are changed to reflect the operation's outcome. CY is restarted. The AC is on. For instance, ANA B or ANA M

Logical AND immediate with accumulator

ANI   8-bit data   8-bit data for ANI The accumulator's contents and the 8-bit data (operand) are logically ANDed, and the result is added to the accumulator. S, Z, and P are changed to reflect the operation's outcome. CY is restarted. The AC is on. Instance: ANI 86H

Exclusive OR register or memory with accumulator

X RA   R   Following an Exclusive OR operation using the M contents of the operand (register or memory), the result is added to the accumulator's M contents. If the operand is a memory location, the HL registers' contents serve as the operand's address. S, Z, and P are changed to reflect the operation's outcome. AC and CY are restarted. For instance, XRA B or XRA M

Exclusive OR immediate with accumulator

XRI   8-bit data   The 8-bit data (operand) is Exclusive ORed with the contents of the accumulator, and the result is then added to the accumulator. S, Z, and P are changed to reflect the operation's outcome. CY and AC are set to zero. Instance: XRI 86H

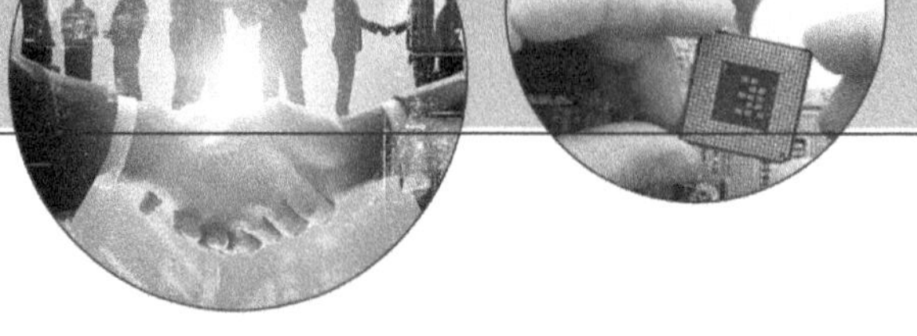

Logical OR register or memory with accumulator

ORA   R   The contents of the register or memory are logically M ANDed with the contents of the accumulator, and the outcome is added to the accumulator. The information in the HL registers determines the operand's address if it is a memory location. To reflect the outcome of the procedure, S, Z, and P are adjusted. Resetting CY and AC.ORA B or ORA M, for instance

Logical OR immediate with accumulator

ORI   Data in ORI 8-bit The accumulator's contents andthe 8-bit data (operand) are logically ORed, and the result is then added to the accumulator. S, Z, and P are changed to reflect the operation's outcome. AC and CY are restarted. Instance: ORI 86H

Rotate accumulator left

RLC   none   The accumulator's binary bits are all rotated one position to the left. The Carry flag and bit D7 are both positioned in the D0 and D7 positions, respectively. In accordance with bit D7, CY is altered. Not impacted are S, Z, P, and AC. Instance: RLC

Rotate accumulator right

RRC   none   The accumulator's binary bits are each rotated right one position. The Carry flag and bit D0 are both positioned in the D7 location. CY is changedS, Z, P, and AC remain unaffected, as indicated by bit D0.Examples: RRC

Rotate accumulator left through carry

RAL   none   Through the Carry flag, each binary bit of the accumulator is rotated leftward by one position. The Carry flag is updated in accordance with bit D7, and the Carry flag is set in the least significant position D0. AC, S, Z, and P are unaffected. Instance: RAL

Rotate accumulator right through carry

RAR   none   Through the Carry flag, each binary bit of the accumulator is rotated rightward by one position. CY is changed in accordance with bit D0, which is placed in the Carry flag, which is placed in the most significant location D7. AC, S, Z, and P are unaffected.Consider RAR.

Complement accumulator

CMA   none   Complementary material is added to the accumuator's contents. Flags are unaffected by this.CMA, for instance.

Complement carry

CMC   none   The complement is the Carry flag. Other flags are not impacted.as in CMC

Set Carry

STC   none   The has the Carry flag set to 1. None of the other flags are impacted. Consider STC.

22.3.5 Control Instructions

Opcode Operand Description

No operation

NOP   none   No procedure is carried out. The command is retrieved and decoded. However, nothing is actually done. Typical: NOP

Halt and enter wait state

HLT   none   The CPU completes the currently running instruction and stops any additional execution. To get out of the standstill state, you need to interrupt or reset. Instance: HLT

Disable interrupts

DI   none   All interrupts other than the TRAP are disabled, and the interrupt enabling flip-flop is reset. Flags are unaffected by this. Instance: DI

Enable interrupts

EI   none   All interrupts are enabled and the flip-flop controlling interrupt enable is set. Flags are not impacted. The interrupt enable flipflop is reset upon a system reset or the acknowledgment of an interrupt, which disables the interrupts. The interrupts must be reenabled using this instruction, with the exception of TRAP. Instance: EI

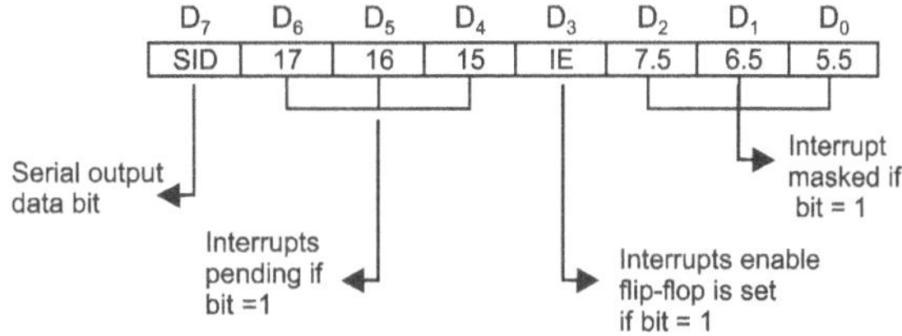

Read interrupt mask

RIM   none   This instruction has multiple purposes, including reading the state of interrupts 7.5, 6.5, and 5.5 and reading the serial data input bit. The following interpretations correspond to the instruction's loading of eight bits into the accumulator. Consider RIM.

Set interrupt mask

None SIM   This general-purpose instruction enables the 8085 processor's interrupts 7.5, 6.5, and 5.5 as well as serial data output. The

following

is how the instruction interprets the contents of the accumulator. Consider SIM.

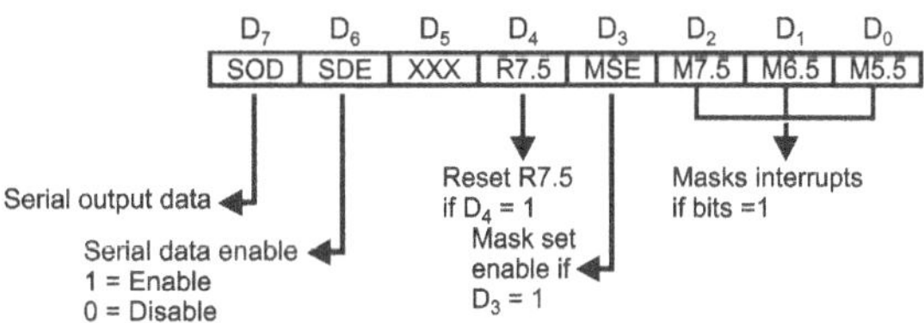

The way an instruction addresses memory or hardware registers can be used to classify it.

SOD-Serial Output Data: If bit D6=1, the accumulator's bit D7 is latching into the SOD output line and made available to a serial peripheral.

SDE - Serial Data Enable: If bit 1 is set, the serial output is enabled. This bit needs to be set to enable serial output.

Do Not CareIf this bit equals 1, the R7.5-Reset RST 7.5 flip-flop is reset. RST 7.5 can be reset with this extra control.

MSE, or Mask Set Enable, is a bit that enables the operations of bits D2+D1+D0 when it is high. Over all interrupt masking bits, this serves as a master control. If this bit is low, the masks are unaffected by bits D2+ and D0.

M7.5-D2 = 0 RST 7.5 is enabled. = 1 RST 7.5 is hidden or turned off.

M6.5-D0 = 0 RST 6.5 is enabled. = 1 RST 6.5 is muted or turned off.

M5.5-D0 = 0 RST 5.5 is enabled. = 1 RST 5.5 is muted or turned off.

22.3.5.1 Implied Addressing:

The function of some instructions assumes a given addressing mode. For instance, the DAA (decimal adjust accumulator) instruction deals with the accumulator while the STC (set carry flag) instruction solely deals with the carry flag.

22.3.5.2 Register Addressing:

Implied Addressing: The function of some instructions assumes a given addressing mode. For instance, the DAA (decimal adjust accumulator) instruction deals with the accumulator while the STC (set carry flag) instruction solely deals with the carry flag.Register Addressing is required for a sizable number of instructions. These instructions require the operation code and one of the registers A through E, H, or L. The accumulator is inferred as a second operand in these instructions. Compare the contents of the E register with the contents of the accumulator, for instance, according to the instruction CMP E.

Register addressing instructions frequently work with 8-bit values. Some of these instructions, though, work with 16-bit register pairs. For instance, the PCHL instruction swaps the contents of the H and L registers with those of the programme counter.

22.3.5.3 Immediate Addressing:

When immediate addressing is used, the data that has to be assembled is already included in the command. As an illustration, the instruction CPI 'C' could be read as 'compare the contents of the accumulator with the letter C. The hexadecimal value of this instruction after assembly is FE43. The internal representation of the letter C is 43 in hexadecimal. The CPU fetches the first instruction byte when this instruction is executed and then realises it needs to fetch one more byte. The following byte is fetched by the CPU and placed in one of its internal registers before the compare operation is carried out.

It's important to note that the titles of the instant instructions make it clear that they employ immediate data. As a result, an add instruction is designated as ADD, while an add immediate instruction is designated as ADI.

The accumulator is used as an inferred operand in all but two of the instantaneous instructions, much like in the CPI instruction previously demonstrated.

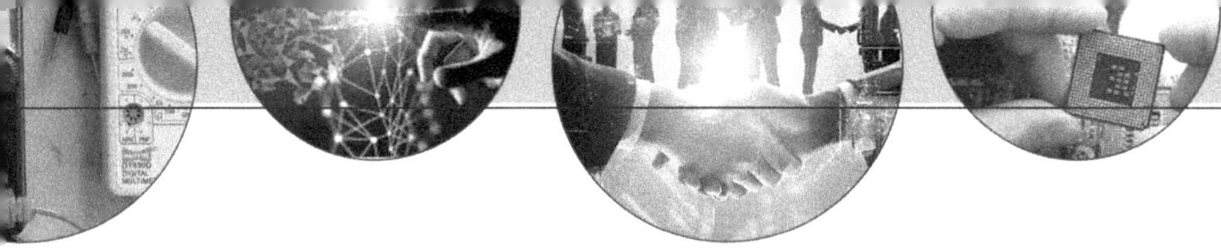

The MVI (move instant) instruction has the ability to move the immediate data it contains to memory or any other working register, including the accumulator. Thus, the hexadecimal value FF is sent to the D register by the instruction MVI D, OFFH.

Even more peculiarly, the immediate data of the LXI instruction (load register pair immediate) is a 16-bit value. A register pair is frequently loaded with addresses using this instruction. The stack pointer must be initialised in your programme, as was previously explained. The most popular instruction for this is LXI. For instance, the stack pointer is loaded with the hexadecimal value 30FF by the instruction LXI SP,3OFFH.

22.3.5.4 Direct Addressing:

: A 16-bit address is part of the instruction for jump instructions. By changing the contents of the programme counter with the new value 1000H, for instance, the instruction JMP 1000H triggers a jump to the hexadecimal location 1000.

Three bytes of storage are needed for instructions with direct addresses: one for the instruction code and two for the 16-bit address.

22.4 8085 SAMPLE PROGRAMS

22.4.1 EXPERIMENT NO. 1

To study the data transfer

Programmed Data

Memory Address			Opcode Operand	Hexcode	Remark
2000	MVI	A, 20	3E	20, is moved to ACC.	
2001	20				
2002	MVI	B, 00	06	Clear Register B	
2003	00				
2004	MOV	B, A	47	The contents of ACC. are moved in Reg. B	
2005	STOP	-	76	Stop	

The procedures for starting a programme and running it
1. Click the RESET key.
2. Hit the REL key.
3. Type the beginning address, for instance 2000.
4. To enter data (Hex Code), press the NEXT key.
5. Continue Step 4 up to 76 times to finish entering the programme.
6. Turn on the microprocessor trainer kit in order to run the programme.
7. Press the RESET button.
8. Click the SI key.
9. Type your starting address, such as 2000.
10. Press the NEXT key up to the last address, 76.
11. Hit the FILL key.
Press the SHIFT + SI key.
13. Click Reg. (e.g. B) to view the outcome.

22.4.2   EXPERIMENT NO. 2

To study the data transfer group

Programmed Data

Memory Address	Opcode	Operand	Hexcode	Remark
2000	MVI	A, 40	3E	40, is moved to ACC.
2001			40	
2002	MVI	C, 00	0E	Clear Register C
2003			00	
2004	MOV	C, A	4F	The contents of ACC. are moved in Reg. C
2005	HLT	-	76	Stop

22.4.3   EXPERIMENT NO.3

To study the data subtraction

Programmed Data

Memory Address	Opcode	Operand	Hexcode	Remark
2000	MVI	A, 40H	3E	40, is moved to ACC.
2001			40	
2002	MVI	B, 20H	06	20,data is moved in Reg.B
2003			20	
2004	SUB	B	90	Subtract the content of Reg.B from ACC.
2005	MOV	C,A	4F	Move the result in the specified Register
2006	HLT	-	76	Stop

22.4.4   EXPERIMENT NO.4

Mark D1 to D4 to load A and H into Reg. C. Display lower order at a memory location (or some Reg.

Programmed Data

Memory Address	Opcode	Operand	Hexcode	Remark
2000	MVI	C,AH	3E	Load A&H in Reg.C
2001			AH	
2002	MOV	A,C	79	Bringing A&H in Acc.
2003	ANI	0FH	E6	Mask bits D1 to D2
2004		0F		
2005	STA	Xx50	32	Display low order
2006			50	at memory location xx50
2007			xx	
2008	MOV	B,A	47	Display low order at Reg.B
2009	HLT	-	76	Stop

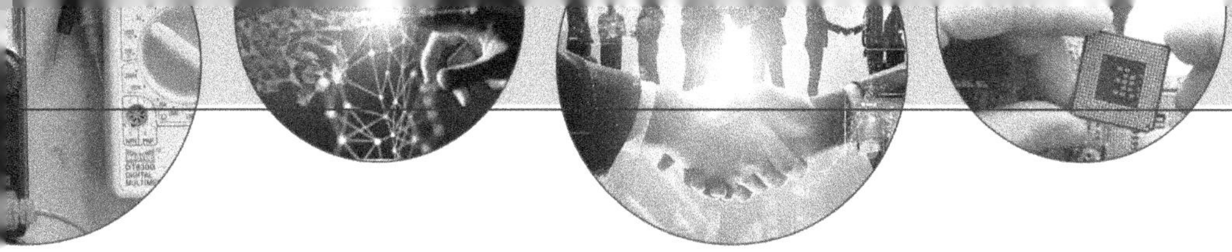

22.4.5   EXPERIMENT NO.5

A programme that adds two 8-bit numbers together produces an 8-bit number as a result.

Programmed Data

Memory Address	Opcode	Operand	Hexcode	Remark
2000	MVI	A, 40H	3E	40, is moved to ACC.
2001			40	
2002	MVI	B, 20H	06	20,data is moved into
2003			20	Reg.B
2004	ADD	B	80	Add the content of Reg.B from ACC.
2005	MOV	C,A	4F	Move the result in the specified Register
2006	HLT	-	76	Stop

MS nybble LS nybble	h'0' b'000'	h'1' b'001'	h'2' b'010'	h'3' b'011'	h'4' b'100'	h'5' b'101'	h'6' b'110'	h'7' b'111'
h'0' b'0000'	NUL	DLE	SP	0	@	P	'	p
h'1' b'0001'	SOH	XON	!	1	A	Q	a	q
h'2' b'0010'	STX	Dc2	"	2	B	R	b	r
h'3' b'0011'	ETX	XOFF	#	3	C	S	c	s
h'4' b'0100'	EOT	Dc4	$	4	D	T	d	t
h'5' b'0101'	ENQ	NAK	%	5	E	U	e	u
h'6' b'0110'	ACK	SYN	&	6	F	V	f	v
h'7' b'0111'	BEL	ETB	'	7	G	W	g	w
h'8' b'1000'	BS	CAN	(	8	H	X	h	x
h'9' b'1001'	HT	EM	)	9	I	Y	I	y
h'A' b'1010'	LF	SUB	*	:	J	Z	j	z
h'B' b'1011'	VT	ESC	+	;	K	[	k	{
h'C' b'1100'	FF	FS	,	<	L	\	l	|
h'D' b'1101'	CR	GS	-	=	M	]	m	}
h'E' b'1110'	SO	RS	.	>	N	^	n	~
h'F' b1111'	SI	US	/	?	O	_	o	DEL

Table: ASCII, the American Standard Code for Information Interchange, is a 7-bit system

---

CHAPTER-23

Interfacing with Microprocessor

23.1 8085 MICROPROCESSOR INTERFACING

An external crystal or RC network is necessary for the internal clock generator. At twice the standard CPU operating frequency, it will oscillate. This oscillator internally generates a two-phase, non overlapping clock with a 50% duty cycle, and one phase of the clock (2) is available as an external clock. The external RDY synchronisation that the 8224 (Clock generator) formerly offered is now directly provided by the 8085A. Schmitt action input is built into the RESET IN input, requiring only a resistor and capacitor for power-on reset. System RESET is supported by RESET OUT.

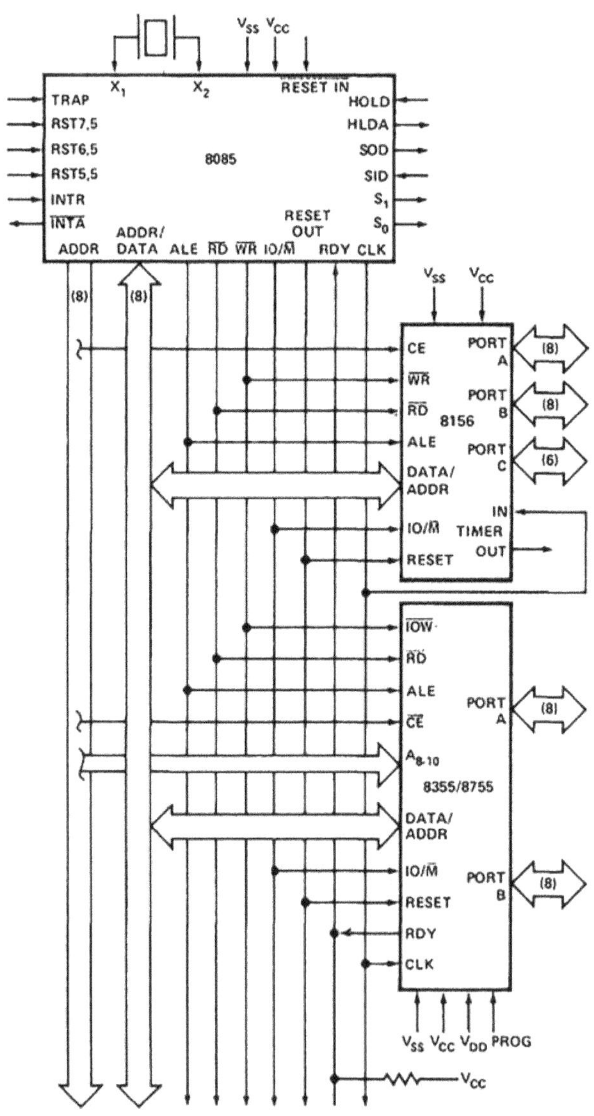

fig. 23.1 : 8085 Microprocessor Interfacing

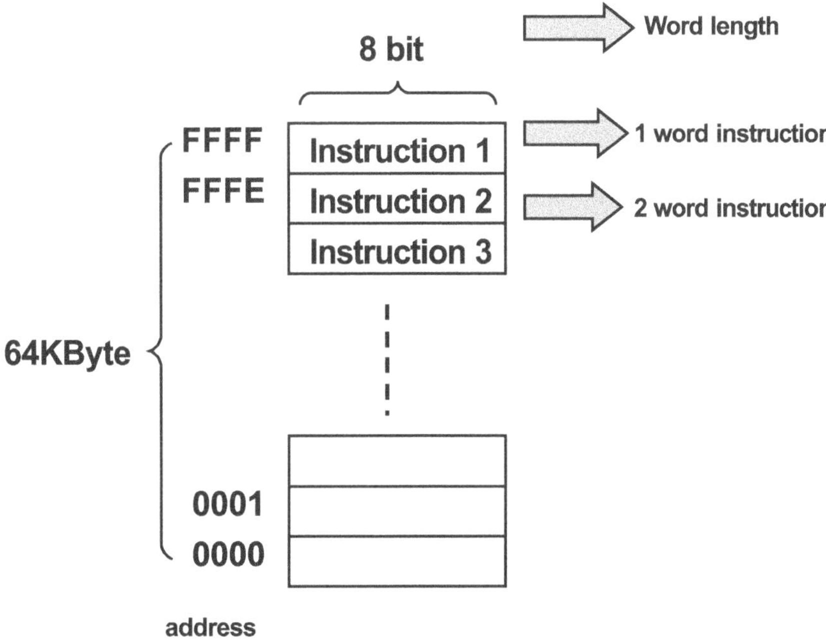

fig. 23.2 : 8085 Microprocessor memory interface

According to the memory interface block diagram in Fig. 23.2:

Word: the range of bits (4–64bit) that the micro-P can recognise and process at once. **Instruction:** a combination of bit patterns with known-to-the-micro-P particular meanings. **Programme:** A complete set of instructions. Situation Reports Direct access to status information is provided by the 8085A.

23.1.1 ALE acts as a status indicator. Since the status is only partially encoded, the user can determine the type of bus transfer being done ahead of time.

Additionally, a direct IO/M cycle status signal is provided. Decoded Therefore, S carries the following status

23.1.2 Data: In all bus transfers, HALT WRITE READ FETCH S1 can be understood as R/W. The 8 LSB of the address are multiplexed with the data in the 8085A rather than the status.

The lower half of the address is entered into the memory or peripheral address latch using the ALE line as a strobe. As a result, more pins are available for increased interrupt capabilities.

23.1.3 Interrupt and Serial I/O Five interrupt inputs are available on the 8085A: INTR, RST5.5, RST6.5, RST7.5, and TRAP. The 8080 INT and INTR both perform the same tasks. Three RESTART inputs—inputs 5.5, 6.5, and 7.5—each include a configurable mask. While TRAP is a RESTART interrupt, it cannot be muted.

The three RESTART interrupts, which save the programme counter in the stack and branch to the RESTART address, induce internal execution of RST if interrupts are enabled but the interrupt mask is not set. Independent of the state of the interrupt enable or masks, the nonmaskable TRAP triggers the internal execution of a RST.

23.1.4 RESTART Name Address (Hex) TRAP RST 5.5, 6.5, and 7.5, 2416, 2C16, 3416, and 3C16 respectively.

The internal flip flop that creates the internal interrupt request for RST 7.5 only has to be set by a pulse. The RST 7.5 request flip flop stays set until the request is fulfilled. After that, it automatically resets. The SIM instruction or a RESET IN command to the 8085A can likewise be used to reset this flip-flop. Even if the RST 7.5 interrupt is disabled, a pulse on the RST 7.5 pin will still set the internal flip flop of the RST 7.5 device.

Only the SIM instruction and RESET IN can modify the status of the three RST interrupt masks.

If there are multiple pending interrupts, the interrupt with the highest priority is the one that is recognised: Priorities are as follows: TRAP highest, RST 7.5, RST 6.5, RST 5.5, and INTR lowest. This priority method does not consider the priority of a procedure that was initiated by an interrupt with a higher priority. If the interrupts were reenabled before the completion of the RST 7.5 function, RST 5.5 may interrupt the RST 7.5 process.

23.2 SERIAL I/O AND INTERRUPTS

A microprocessor's interface to the outside world

- Input: Keyboard, mouse
- Output: a printer and a monitor

23.2.1 System Bus – wires linking memory and I/O to the CPU.

- Address Bus: unidirectional
- Data Bus: bidirectional
- Control Bus: synchronisation signals, timing signals, and control signal.
- Transferring data
- identifies location of peripheral or memory

23.2.2 Fundamentals of Microprocessors

distinctions between:

Microcomputer – A computer with a microprocessor as its central processing unit is referred to as a microcomputer.

Microprocessor – The silicon device known as a microprocessor contains an ALU, register circuitry, and control circuits.

Microcontroller – Microcontrollers are silicon chips that combine a CPU, memory, and I/O into a single unit.

Catastrophic errors like a power outage or a bus issue can benefit from the TRAP interrupt. Although it has the highest priority, the TRAP input is recognised just like any other interrupt. Any flag or mask have no impact on it. Both edges and levels are sensitive to the TRAP input. The TRAP input must go high and stay high in order to be acknowledged, and it won't be acknowledged again until it drops low and then back up to high. This prevents any erroneous triggering brought on by background noise or faulty logic. The TRAP interrupt request circuitry in the 8085A is shown in the accompanying diagram

It should be noted that servicing any interrupt (TRAP, RST 7.5, RST 6.5, RST 5.5, or INTR) stops all upcoming interrupts (apart from TRAPs) until an El command is carried out.

The unique feature of the TRAP interrupt is that it keeps the previous interrupt enable status. It is possible to detect whether interrupts were enabled or disabled before the TRAP by running the first RIM command after the interrupt. The current interrupt enabling status is provided in every RIM instruction that follows. Additionally, the RIM and SIM instructions regulate the serial I/O system. RIM reads SID, and SID creates the SOD data.

Basic System Timing

The Data Bus of the 8085A is multiplexed. To sample the lower 8 bits of address on the Data Bus, ALE is used as a strobe. An instruction fetch, memory read, and l/ O write cycle (OUT) are displayed in Fig. 23.2. The l/O port address is copied on both the upper and bottom halves of the address during the l/O write and read cycle, it should be noted.

23.3   SYSTEM INTERFACE

Memory parts from the 8085A series are directly compatible with 8085A CPUs. For instance, the three chips 8085A, 8156, and 8355 together form the following system:

1 Timer/Counter, 2K Bytes ROM, 256 Bytes RAM, 4 8-bit I/O ports, 1 6-bit I/O port, and 4 interrupt levels. Ports for serial in and out

Using the common l/O approach, this minimal system. The memory mapped l/O provides an effective l/O addressing option in addition to normal l/O technique. With this method, a section of memory's address space is designated for l/O addresses, allowing l/O manipulation to take place. Memory Mapped I/O using 8085A system setup.

The conventional memory, which lacks a multiplexed address/data channel, can also communicate with the 8085A CPU. A straightforward 8212 (8-bit latch) will be needed.

23.4 Intel 8088 microprocessor- clock and reset circuitry

The Intel 8088 Microprocessor will receive a system clock from this design, as shown in fig. 23.3. Additionally, this design will offer a way to reset the Intel 8088 Microprocessor.

The system will be powered by a mains supply of +5V DC.

The 8088 microprocessor will receive a 4 MHz clock from a 12 MHz crystal oscillator with a 33% duty cycle.

The reset circuit will offer a power-on reset in addition to a manual reset using a pushbutton switch.· An always-on READY signal will be sent to the 8088 microprocessor.

System Clock, Ready, and Reset are the three fundamental signals provided by this design to the Intel P8088 Microprocessor. It will use a 4 Mhz clock with a 33% duty cycle as the system clock. This clock will be created using an 8284 clock generator and a 12 MHz crystal oscillator. Since this design does not call for the use of wait states, the 8284 clock generator will deliver a constant state READY output to the P8088 microprocessor.

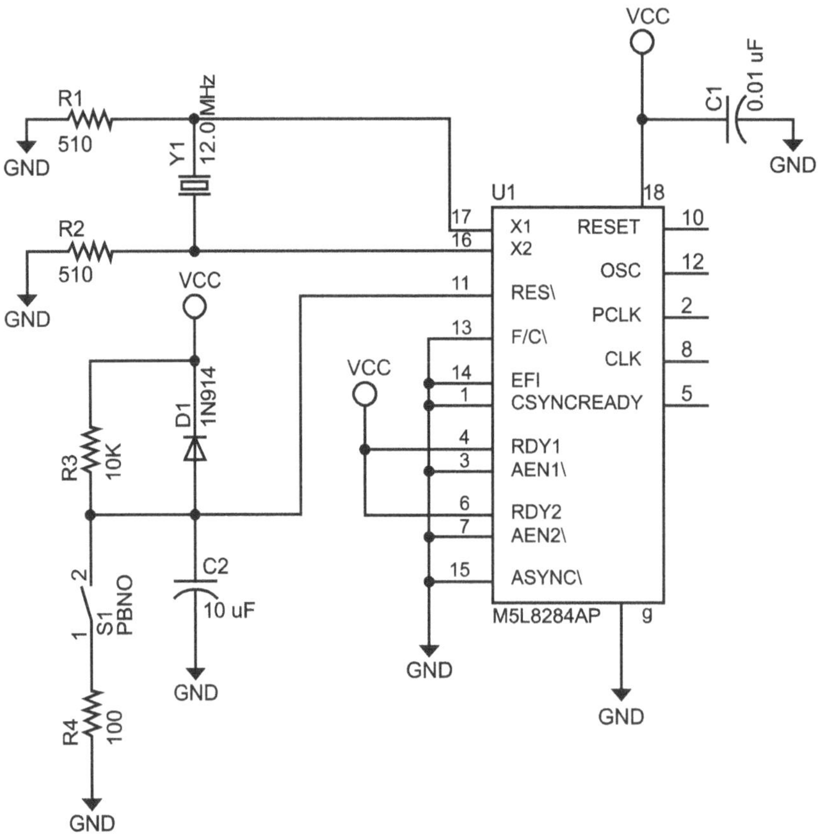

fig. 23.3, Intel 8088 μP-circuitry

A pushbutton switch is used to manually reset the P8088 Microprocessor. The RESET to the P8088 Microprocessor will be asserted during power up until the system power has stabilised.

An 8284 Clock Generator/Driver chip will be used to power the system clock. This clock generator (U1) will generate a 4 MHz system clock using a 12 MHz crystal oscillator (Y1). The Intel P8088 Microprocessor utilised later in this project requires a 33% duty cycle, which the 8284 clock generator offers. The 8284 will be set up so that it will provide an always-true READY signal to the Intel 8088 microprocessor.

The output READY will remain active if the inputs for ready are connected to either VCC or GND as indicated in the schematic diagram in Figure 1. The clock generator U1 is told to use the crystal oscillator Y1 as the source of the clock when the input F/C is grounded. The crystal oscillator Y1 is stabilised by the resistors R1 and R2.

The time constant produced by C2 and R3 is used by the reset circuitry to provide a power-on reset. Due to this time constant, the reset input at U1-11 stays in the true state for 28 mS, allowing the system power to stabilise. A manual way of asserting reset at U1-11 is also provided via S1 and R4. R4 controls S1's current flow to prevent switch deterioration. A lack of system power will cause C2 to discharge through D1, preventing C2 from discharging through U1-11.

The clock rate at the clock generator U1's CLK output can be seen using an oscilloscope.

The READY output of clock generator U1 can be seen using an oscilloscope.

The clock generator U1's RESET output can be seen on an oscilloscope both during power-up and during a manual reset

The clock generator U1-8's output displayed a TTL level 4 MHz clock with a 250 nS period.

The clock's duty cycle was 88 nS, or 35% of its normal duty cycle.A TTL high is produced by the output at U1-5, which is a steady 4.95V DC.

The P8088 microprocessor is held in reset mode by the power-on reset circuitry for 28 mS after the system voltage reaches +5V. The manual reset goes as planned and has no impact on the system's power.

A 15 MHz oscillator might be used to provide the Intel P8088 Microprocessor a 5 MHz clock, which would improve the circuit. By giving the Intel P8088 CPU a 5 MHz system frequency, which it was built to operate at, processor performance would be enhanced.

23.5   8086 AND 8088 MICROPROCESSORS

The i8088 CPU served as the basis for the original IBM PC. Although it has a 4 byte instruction prefetch queue and an 8-bit external data bus (16-bit internally), this device is fully software compatible with the i8086. For our purposes, the 8086 and 8088 can really be viewed as one unit.

The 8086/8088 microprocessor has two core components: the bus interface unit and the execution unit (EU), which fetches instructions, reads operands, and writes results. This is seen in the block diagram of Fig. 23.4. The BIU permits some overlap between the fetching of instructions and their execution.

The internal organisation of the 8086/8088 is depicted in fig. 23.5:

The following components make up the Bus Interface Unit (BIU):

As the processor is working on the current instruction, the next instructions or data can be fetched from memory using the instruction queue. This separates the memory cycle time from the execution time because the memory interface is typically significantly slower than the processor.

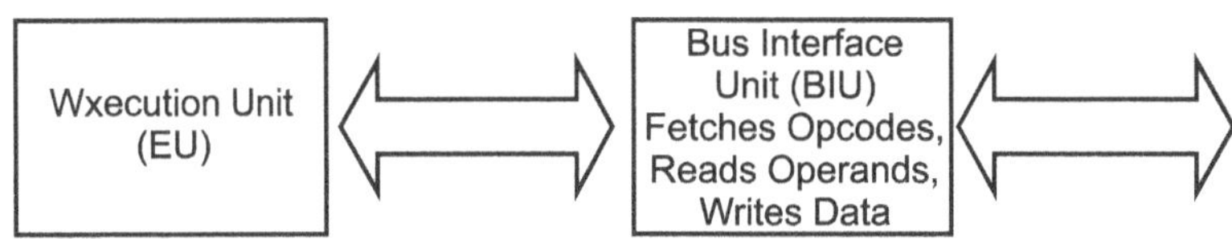

fig. 23.4 : Block diagram of 8086/8088 Microprocessor

23.6 8086/8088 ARCHITECTURE

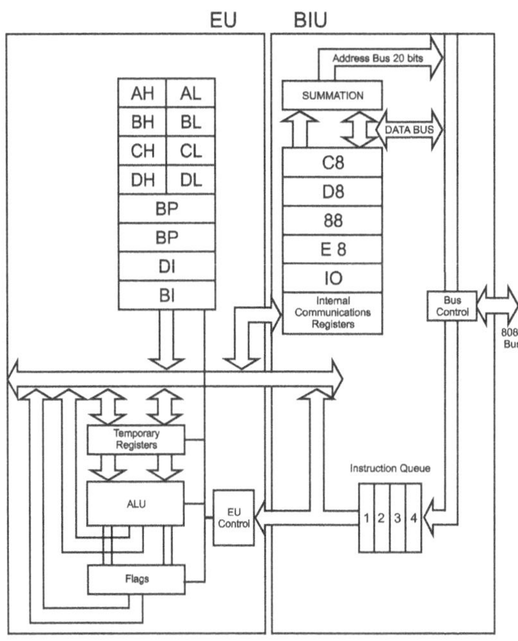

fig. 23.5: Architecture of 8086/8088 Microprocessor

Code Segment (CS), Data Segment (DS), Stack Segment (SS), and Extra Segment Registers

To create the 20-bit address needed for the 8086/8088 to address 1Mb of memory, segment (ES) registers, which are 16-bit registers, are combined with the 16-bit Base registers. As a programme runs, they are changed to point to different segments under programme control. The 8086 uses the Segmented architecture to maintain compatibility with previous CPUs like the 8085. It is among the Intel Architecture's most important components.

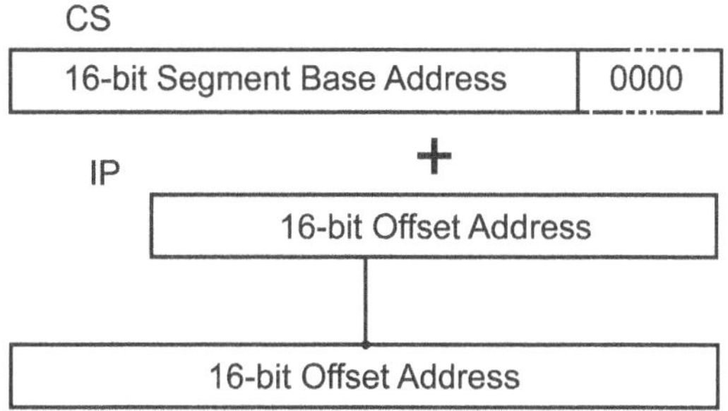

The Address and Instruction Pointer (IP) Summary: The following instruction's Offset Address, which is contained in the IP, is measured in bytes from the base address provided by the current Code Segment (CS) register. How to do this is depicted in the figure.

The CS's contents are moved left by four. Bit 15 is moved to Bit 19's location. There are zeros in each of the bottom four bits. A 20-bit physical address is created by adding the calculated value to the contents of the Instruction Pointer. The IP is viewed as an offset into the segment whose base address is made up by the CS.

For all other general registers and segment registers in the 8086 device, this segmented approach also applies. For instance, the stack region in physical memory is addressed using the same combination of the SS and SP.

23.7   THE 8088 MICROPROCESSOR IN CIRCUIT

To function in a microcomputer system, the 8086 and 8088 microprocessors need some additional support circuitry. These processors, for instance, multiplex the address and data buses on the same pins in order to fit in a 40-pin package. In order to create independent address and data buses that can connect with hardware like RAMs and ROMs, some demultiplexing circuitry is also required. In fact, the 8288 Bus Controller, 8284A Clock Generator, 74HC373 and 74HC245 devices must all be supported in order for the 8086 or 8088 to operate in Maximum Mode.

Be aware that the 8288 rebuilds the microprocessor's normalbus control signals (MRDC#, MWTC#, IORC#, IOWC#, etc.) in Maximum mode using a set of status signals (S0, S1, S2). We'll quickly review a few 8086/8288 pins' functions

RESET: The processor clears the flags register, segments registers, and other registers in addition to setting the effective programme address to FFFF0h (CS=FFFFh, IP=0000).

(8086 only) BHE# The 8086's 16-bit data bus and byte-wide memory are connected via Bus High Enable (BHE#), which is used in conjunction with the A0 signal. How BHE# and A0 are combined is shown in the table:

BHE#	A0	Selection
0	0	Whole word (16-bits)
0	1	High byte to/from odd address
1	0	Low byte to/from even address
1	1	No selection

This is how memory is accessed using these signals

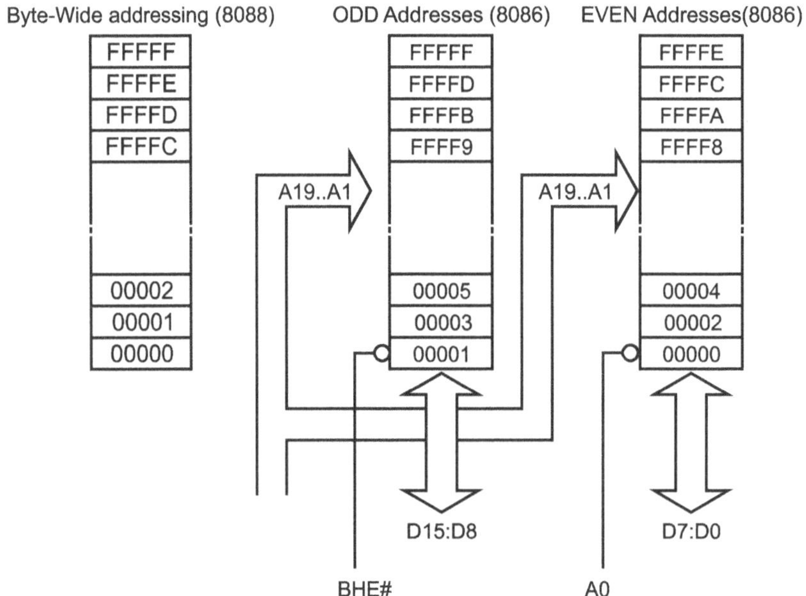

Even when 16-bit memory are used, this technique still holds true. It makes byte data accessible to the 8086. 32-bit CPUs like the 80386 are able to access bytes thanks to similar techniques.

23.8 MANUFACTURERS OF 8085 MICROPROCESSOR

Microprocessor type Beginning: 1976, 8 bus width between 3 and 8 Mhz

The following iteration of the Intel 8080 CPU family is the Intel 8085 microprocessor.

The 8085 was quicker than the 8080 and had the following improvements:

The Intel 8085 included a single 5-volt power supply, an integrated clock oscillator and system controller, a serial I/O interface, and two new instructions that were added to the instruction set. There were a few undocumented instructions on the CPU as well.

These instructions were originally planned to be a component of the CPU instruction set, but they were abruptly removed since they were incompatible with the upcoming Intel 8086.

Differences between Intel 8080 and 8085 processors are fully listed. The 8085 microprocessor came in a number of variations.

The 8085 microprocessor, which had no suffix "A" in the beginning, was solely produced by Intel and was swiftly superseded by the 8085A, which had problem fixes. A short time later, in the early 1980s, Intel unveiled the HMOS version of 8085A, the 8085AH. A CMOS variant of the 8085A, the 80C85A, was also available. Even though it's unclear whether Intel ever produced the 80C85, it was made by at least two other companies, OKI and Tundra Semiconductor.

The 8085 microprocessor that runs at 8 MHz and was produced by Tundra Semiconductor is the fastest. AMD, Mitsubishi, NEC, OKI, Siemens, and Toshiba are second-source manufacturers. Soviet Union also produced imitations of the Intel 8085 CPU.

Manufacturers- AMD
S_AMD-AM8085A-2DC-C8085A-2
5 MHz
40-pin ceramic DIP
Purple ceramic/gold top/gold pins

Manufacturers- Mitsubishi
S_Mitsubishi-M5L8085AP
3 MHz
40-pin plastic DIP

Manufacturers- NEC
S_NEC-D8085A
3 MHz
40-pin ceramic DIP
White ceramic/silver top/gold pins

Manufacturers- OKI
S_OKI-MSM80C85ARS

CMOS-based OKI MSM80C85A microprocessors had a few benefits over NMOS-based 8085 microprocessors. In comparison to NMOS 8085 microprocessors, MSM80C85A microprocessors used less power and had a wider operating temperature range of -40 to 85 °C. In comparison to an Intel 8085 microprocessor operating at the same frequency, the OKI MSM80C85ARS uses 8 times less power.

Manufacturers- Siemens
S_Siemens-SAB8085A-C-silver
3 MHz
40-pin ceramic DIP
Purple ceramic/silver top/tin pins

Manufacturers- Toshiba
S_Toshiba-TMP8085AP
3 MHz
40-pin plastic DIP

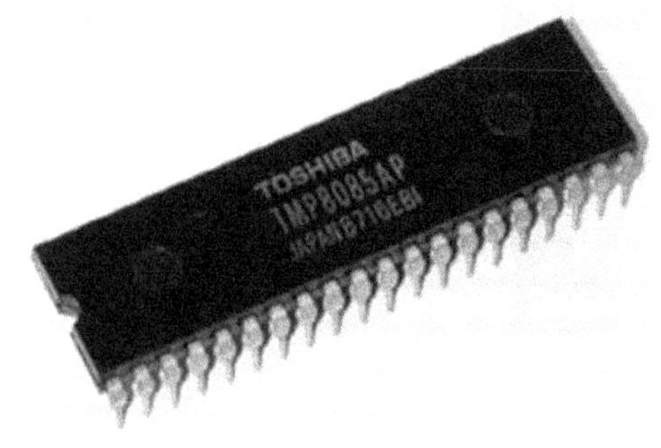

Manufacturers- USSR
S_USSR-IM1821VM85A
3.6 MHz
40-pin ceramic DIP
White ceramic/silver top/tin pins
Soviet clone of Intel 80c85 processor
Military version

210

23.9   APPLICATIONS OF MICROPROCESSOR

Measurement of petrol in a car tank

Let's examine a straightforward application. Think about the setup in Fig. 23.6. Let's examine the task that the processor must complete. Finding out what the task is, or, to use the formal engineering terminology,

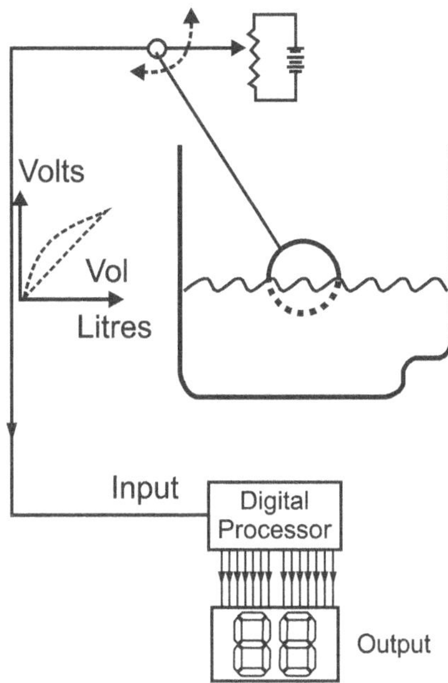

fig. 23.6 : Measuring the quantity of petrol in a car tank

It begins with a system analysis that establishes the requirements, then moves on to the implementation phase that produces the desired outcome.

In this scenario, we want to design the electronics to measure the gasoline capacity of a car with a resolution of one litre and a 2-digit decimal display, measuring litres up to 99. We can divide the system into three parts using the structure depicted in the diagram.

Input

Along with the liquid level, a float at the end of an arm oscillates up and down. This causes a variable resistor (a potentiometer) to rotate, changing the voltage applied to the pot's wiper. This is what auto enthusiasts refer to as a sender.A sensor is an object, such as a potentiometer, that transforms a physical quantity into an electrical variable. The term "transducer" refers to the entire input system, also known as the mechanical arm/float, sensor, and signal processing (such as filtering to reduce noise).

The input quantity is now an analogue voltage that changes quite smoothly. However, because of the geometry of the setup and flaws in the potentiometer, it is not directly

Processor

The processing unit—in this case, a microprocessor—has a number of jobs to do.

• Transform the analogue voltage at the input to its digital equivalent.

• Calculate the equivalent amount of litres by processing the digital input number. For instance, 3.6 volts may equal 001100000 (48 gallons), for example.

• To activate the two lots of 7-segment displays, convert the litre digital quantity to a form.

If a segment is illuminated by a 1, then in this instance 1100110 11111111 for.

Output Use one display per digit as there are two decimal digits to be shown here. These 7-segment displays are displayed. Any decimal number can be formed by lighting the appropriate portions, for example:. If you blend upper and lower case letters, we can also do some letters.

Binary (off/on) digital output is still used to drive the displays. The CPU would need to do a digital to analogue current conversion if we intended to use an analogue metre.

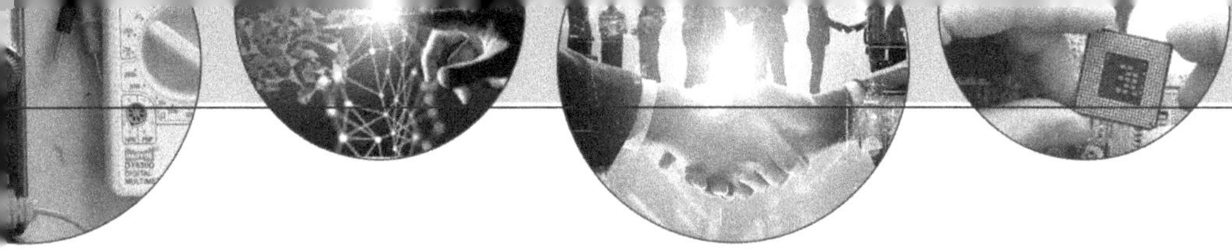

ADCHN FINAL EXAM SAMPLES
PART - I
ANALOG ELECTRONICS

Q.1   How, in terms of electric charges, is "static cling" explained?

Q.2   If an electric field exists to the right, in which direction would a positive charge move if placed in that field?

Q.3   For an Ohmic material, decreasing the potential difference applied across it will cause the current to do what?

Q.4   If electricity costs Rs. 5/- per kilowatt-hour, how much would it cost to run a 1000 Watt heater for 10 hours?

Q.5   What is one way that electric charge and magnetic "charge" are similar? What is one way that they differ?

Q.6   What fundamental property of electricity and magnetism is behind the operation of the following devices:

- electric motor

- speaker

- electric generator

- microphone

- magnetic tape recorder

- magnetic tape reader

Q.7   Why, in an electromagnet, is a ferromagnetic core often found?

Q.8   What is one basic electric and one basic magnetic property of a superconductor?

Q.9.   Find the current in a circuit of equivalent resistance equal to 5 Kohm for different voltages.

(a) 20V

(b) 40V

(C) 60V

Q.10   Find the resistance of a circuit if :
(a) current = 40mA & Voltage = 20 Volts
(b) current = 600mA & Voltage = 30 Volts
(c) current = 20mA & Voltage = 60 volts

Q.11   If you know that the circuit has a 2 amp current flow, use Kirchoff's voltage law and Ohm's equation to determine the value of the unknown resistance R.

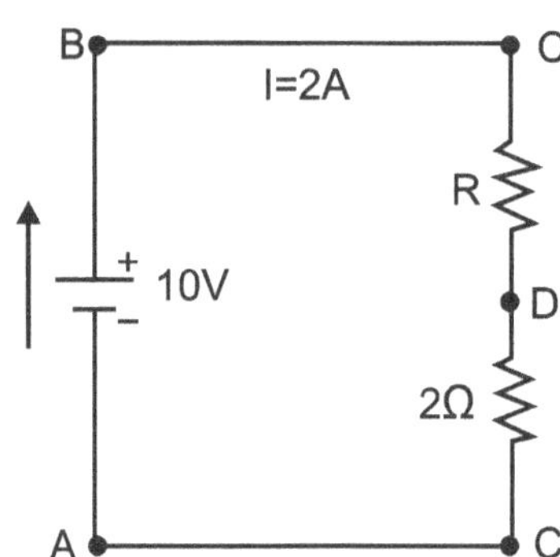

Q.12 If the turn ratio of a transformer is 2:1 and output voltage is 100V, calculate the input voltage.

Q.13 If the rating of transformer is 200 watts and 400 volts is applied at the input and the output current rating is 4 amp. Find the input current and output voltage.

Q.14 Match the pairs :

Part-A	Part-B
Step-up transformer	Turn Ratio<1
Isolation transformer	Turn Ratio>1
Step-down transformer	Oil Cooled
Natural cooled	Turn Ratio=1
Forced cooled	Air cooled

Q.15 Fill in the blanks:

(a) The two most commonly used semiconductor elements are .... and ....

(b) An atom which have 4 valence electrons is called as ....

(c) The semiconductor materials that are free from impurities are referred to as ....

(d) The conductivity of pure semiconductors can be increased considerably by a process called...

(e) Trivalent atoms are referred to as..

(f) Pentavalent atoms are referred to as ....

Q.16 Match the Pairs :

Part-A	Part-B
Forward Bias	Low conductivity
Reverse Bias	Trivalent
Arsenic (As)	High Resistance
Gallium (Ga)	Pentavalent
Intrinsic	Low Res

Q.17 Determine the biasing of the two PN junctions by looking at the energy diagram for a BJT operating in its conducting mode (current flowing through each of the three terminals: emitter, base, and collector):

• Is the emitter-base connection biased forward or backward?

• Is the base-collector connection biased forward or backward? While the transistor is conducting, one of these two junctions actually operates in the reverse-bias mode.

Explain how a basic PN junction makes this possible.

When operating in reverse-bias mode, a diode conducts very little current.

Ans. The base-collector junction is reverse-biased, whereas the emitter-base junction is forward-biased. Injected charge carriers from the emitter enable collector current to flow across the base-collector junction.

Q.18. Match the schematic symbols on the following images of bipolar transistors:

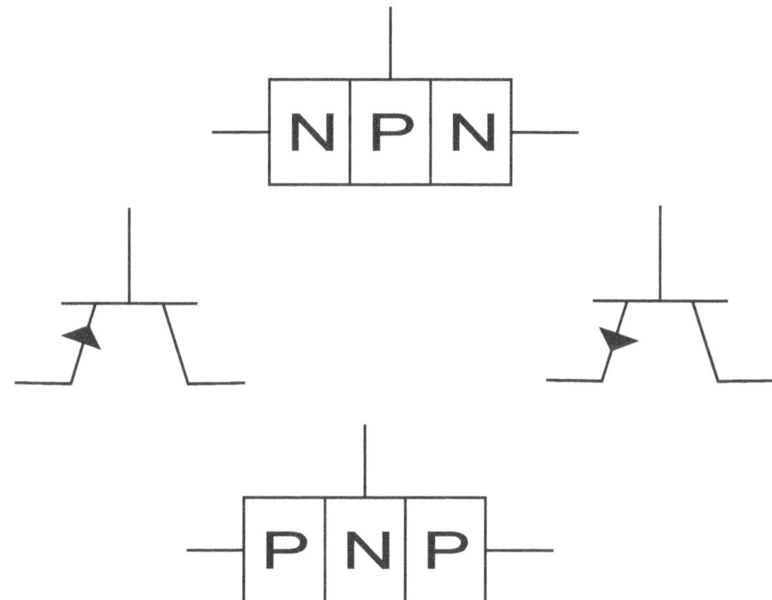

The voltage drop across RB can be used to calculate the different distinct values of IB.

Ans.10 Continued: Name the base, emitter, and collector terminals on each transistor schematic symbol.

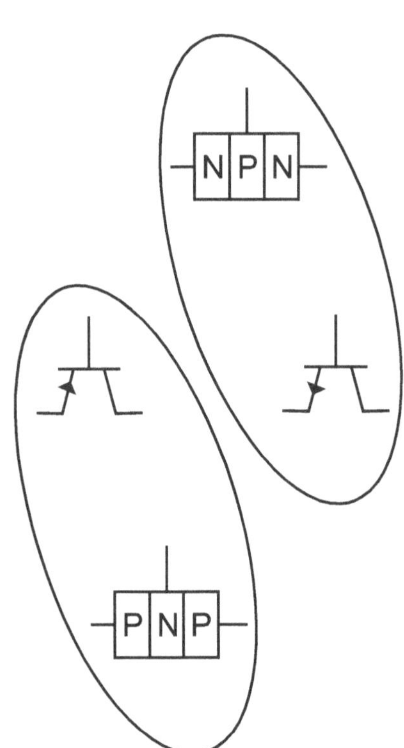

Q.19   Minority carrier devices include bipolar junction transistors. Tell us why?

Ans.   Charge carriers must be "injected" into the transistor's base layer in order for a BJT to conduct, and these charge carriers are always of the "minority" kind in terms of the base's doping.

Q.20   The relative sizes of each current in this bipolar transistor circuit are as follows: What is the difference between the two currents? Are there any currents that, in terms of magnitude, are smaller than the third? Which currents, if any, are they?

Ans. $I_E > I_C \gg I_B$

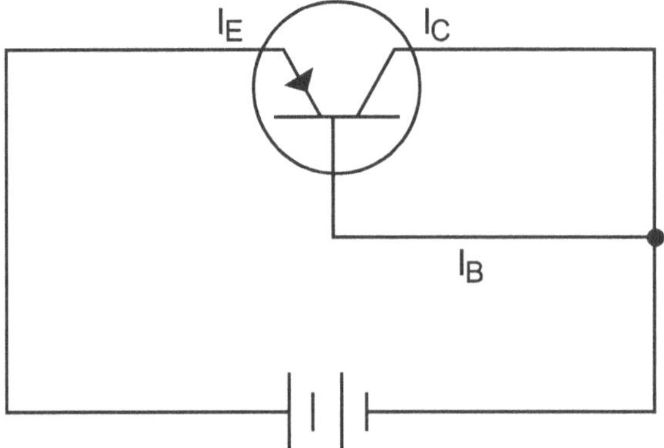

Q.21 Transistors serve as managed sources of current. With a fixed control signal applied, they will typically manage the amount of current flowing through them. Create a test circuit to demonstrate this transistor behaviour. How would you prove that this behaviour of the current regulates itself, in other words?

Ans. Procedure: measure the voltage dropped across R_c while varying V_{cc}.

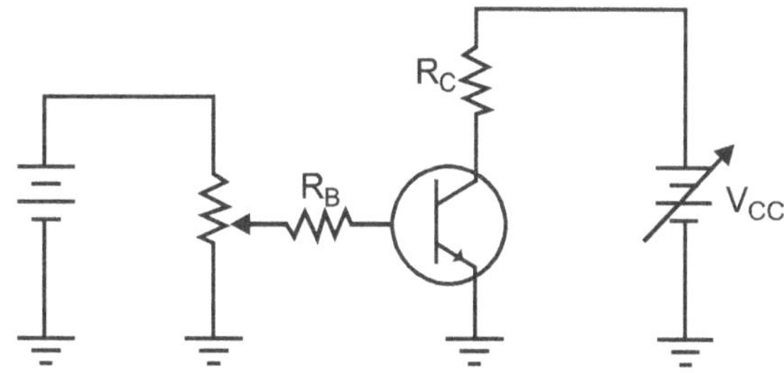

Q.22 Are a transistor's collector and emitter terminals interchangeable? If not, what structural distinction exists between the emitter and the collector?

Ans. Comparatively speaking, the emitter is smaller and more "doped" than the collector.

Q23 Q23 What does the term "open circuit" mean?

Ans. "Open circuit" refers to a lack of connectivity. It's typically used to describe a circuit break, which could be accidental (like a switch left in the open or off position) or intentional (like a frayed wire or burned-out component).

Q24 What is a "short circuit"?

Ans. A "short circuit" is a link that has nearly little resistance, like a wire, and offers a very simple path for electricity. Consider it a shortcut in the electrical system. Instead of a planned connection, it is typically used to describe a flaw or an accidental link.

For instance, we say a battery has experienced a short circuit if the leads from two batteries come into contact and form a connection with a very low resistance. Instead of passing the correct circuit, current will travel through this short circuit. This prevents the circuit from working and could ignite a fire since a sizable current is flowing through the leads and batteries.

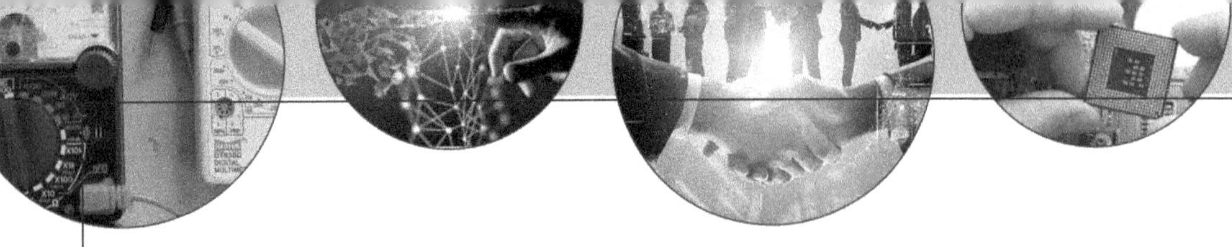

Q.1   a) Write the following decimal numbers as binary numbers to 12-bit accuracy
(i) 1562.5 (ii) 348.678 (iii) 74.8063

     b) Write the following hexadecimal numbers as decimal numbers
(i) 7BA (ii) A25C(iii)C2.57

     c) Write the following decimal numbers as BCD coded binary numbers
(i) 26 (ii)527

     d) Write the Gray-coded equivalent of decimal 37.

     e) Convert the Gray-coded number 1001101101 to its binary equivalent

Q.2   a) Prove the Boolean expression using Venn diagrams

     b) Prove the Boolean expression using truth tables

     c) Verify using Boolean algebra.

Q.3   A modulo-16 ripple counter has a clock frequency of 3.33MHz and the intrinsic timing delay for each Flip-Flop is 75ns.

a)   Draw an accurate timing diagram for this circuit, consider the first 5 full clock pulses only.

     b) Add to your timing diagram the output from the ripple counter (as a decimal number) for each change in the output.

Q.4   A sequential logic circuit is constructed as below the output of the circuit, F, is fed back to form an internal input F' (note the use of the invert bubbles on the inputs to the AND gate)

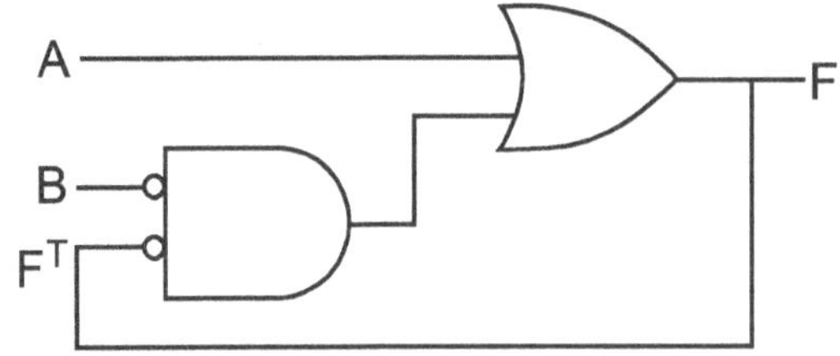

     a) Give the Boolean expression for this circuit.

     b) Create the circuit's truth table.

     c) Indicate the stability of each row in the truth table.

     d) Summarise the behaviour of the circuit with a flow table.

     e) Describe what happens to the circuit as the inputs switch from AB=11 to AB=01.

     f) What is happening to the output F when AB=00?

Q.5   Explain the Binary Number System, what are the advantages of binary code?

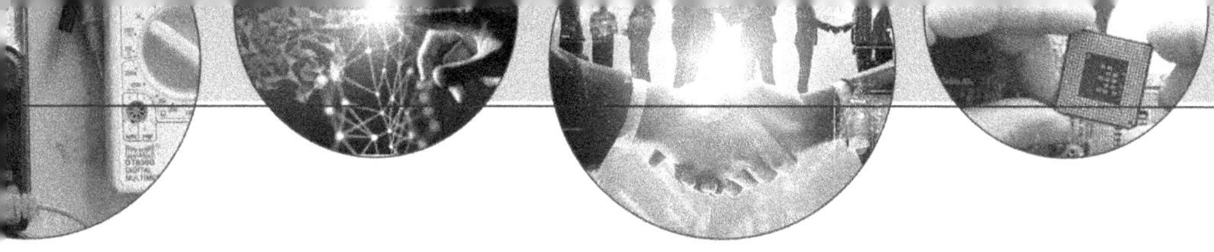

Q.6   Define AND and OR gates. Describe the principles of a decade counter.

Q.7   Draw the Symbolic representation and truth table of following digital gates :

(i)   NAND   (ii) NOR   (iii) NOT

Q.8   Prove de Morgan's theorems
$$\overline{A}.\overline{B} = \overline{A+B}$$
and
$$\overline{A} + \overline{B} = \overline{A.B}$$

with the use of truth tables.

Q.9   Give a truth table for a 4-input NOR gate using the "don't care" condition, where appropriate, to simplify the truth table.

Q.10   The Boolean Expression for the output, Y, from a logic circuit is given by :
$$\overline{Y} = A.B + A.C$$
(i) Draw this circuit.
(ii) Give the truth table for this circuit.

Q11.   What would the output pulse train at the output, Y, look like if the input at C is always 1?

Q.12   Given an OR gate and inverters (NOT gates), draw a logic diagram that will perform the 3-input NAND function.

Q.13   A circuit with inputs A, B, C and D is to be designed such that its output Y is given according to the following truth table.

A	B	C	D	Y	A	B	C	D	Y
0	0	0	0	0	1	0	0	0	0
0	0	0	1	1	1	0	0	1	1
0	0	1	0	1	1	0	1	0	1
0	0	1	1	1	1	0	1	1	1
0	1	0	0	1	1	1	0	0	0
0	1	0	1	1	1	1	0	1	1
0	1	1	0	0	1	1	1	0	0
0	1	1	1	0	1	1	1	1	0

state the simplified minterm expression for this circuit.

Q.14 Choose the correct answers in the following questions.

A) What is the difference between digital and analog techniques?
i) Digital quantities can take on any value over a continuous range.
ii) Digital quantities can take on discrete value over a range.
iii) Actually, they are indifferent, only digital is a new technology invented in 1980s.
iv) None of the above.

B) Click the one which involves digital quantities:
i) Ten-Position switch.
ii) Current meter
iii) Temperature.
iv) Radio volume control in 80s.

C) Which following is not an advantage of digital technique?
i) Digital system is easier to design.
ii) Accuracy and precision are greater.
iii) Digital circuits are less affected by noise
iv) Digital quantities are equivalent to real-world physical quantities.

D) What is the largest decimal number that can be represented using 8 bits?
i) 128
ii) 255.
iii) 256
iv) 1024.

E) Which of the fallowing range is the not used in voltage assignment in digital system:
i) 0.4V - 1.2V
ii) 0.8V - 2V
iii) 0.8V - 2.4V
iv) 1V - 2.4V

F) Convert $(63.25)_{10}$ to binary.
i) 11111.11
ii) 111001.01
iii) 111111.01
iv) 111111.1
v) NA

G) Convert $(1001011.011)_2$ to decimal.
i) 73.0375
ii) 75.375
iii) 91.375
iv) 75.573
v) NA

H) Convert $(11001.1)_2$ to base 8.
  i) $(62.4)_8$
  ii) $(62.1)_8$
  iii) $(31.1)_8$
  iv) $(31.2)_8$
  v) $(31.4)_8$

I) Convert $(25.6)_8$ to binary.
  i) $(10101.11)_2$
  ii) $(11101.10)_2$
  iii) $(10101.10)_2$
  iv) $(10010.11)_2$
  v) $(11111.01)_2$

J) Convert $(35.1)_8$ to base 16.
  i) $(17.4)_{16}$
  ii) $(1D.1)_{16}$
  iii) $(D1.2)_{16}$
  iv) $(E8.1)_{16}$
  v) NA

K) Convert $(39.A)_{16}$ to base 8.
  i) $(35.5)_8$
  ii) $(70.5)_8$
  iii) $(71.5)_8$
  iv) $(72.25)_8$
  v) $(75.5)_8$

L) Convert $(485)_{10}$ to base 16.
  i) $(1E5)_{16}$
  ii) $(231)_{16}$
  iii) $(5E1)_{16}$
  iv) $(15E)_{16}$
  v) NA

M) Boolean algebra is different from ordinary algebra in which way?
  i) More than one discrete level between 0 and 1 can be represented using boolean algebra
  ii) There are only 2 discrete levels in boolean algebra: 0 and 1
  iii) Up to three levels of logic can be described by boolean algebra.
  iv) They are identical in fact
  v) NA The two questions that follow are in reference to the picture below:

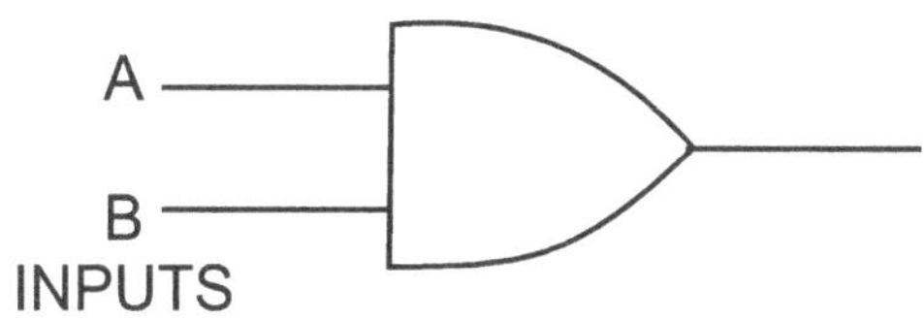

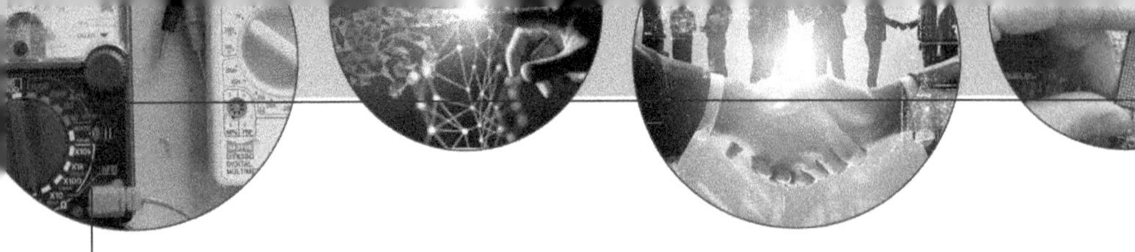

N)   What is the output Y if both inputs A and B are 0?

I) 0

ii) 1

iii) I don't know

iv) NA

O)   If A = 1 and B = 0, what is the output X?
I) 0
ii) 1
iii) I don't know
iv) NA

P)   What inputs are required for a three-input (A,B,C) OR gate if the output is 0?
I) A=0, B=0, C=1
ii) A=0, B=1, C=0
iii) A=1, B=1, C=1
iv) A=0, B=0,C=0
v) NA

The two questions that follow are in reference to the picture below:

Q)   If the inputs are A=1, B=0, and C=1, what is the output X?
i) 0
ii) 1
iii)I don't know
iv) NA

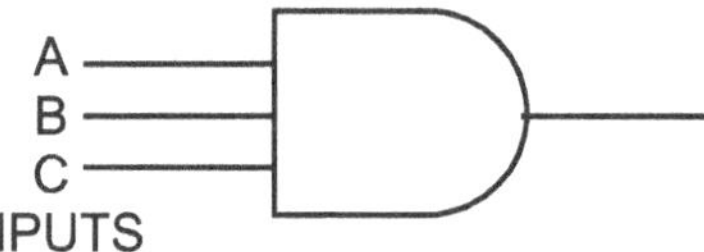

R)   What inputs are needed if output=1?

I) A=0,B=0,C=0
ii) A=1, B=0, C=1
iii) A=0, B=1, C=0
iv) A=1, B=1, C=1
v) NA

Related to the image below are the following two queries:

S)   If the above gate's inputs are 0 and 1, what is the output?
I) 0
ii) 1
iii) Not sure
iv) NA

T)   If output = 1, what are the input values?
I) A=0, B=0
ii) A=0, B=1
iii) A=1, B=0
iv) A=1, B=1
v) I don't know

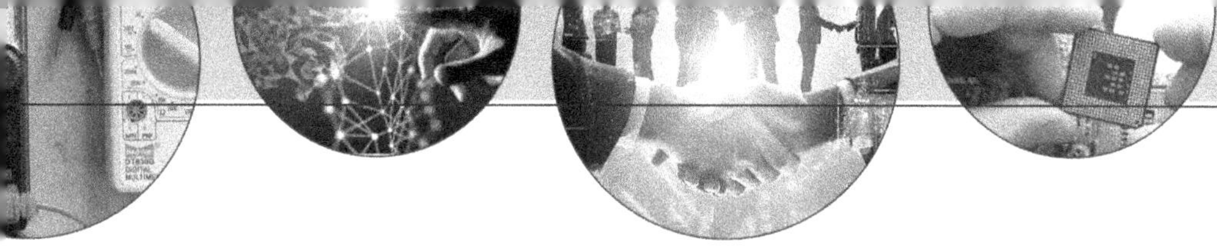

Q.15 Design a full-adder using Logic gates.

Q.16 Explain in detail about IC 74154 with truth table and Pin diagram.

Q.17 Discuss atleast 4 commercial applications of multiplexers & demultiplexers.

Q.18 Write short notes on following :
   a) JK flip-flop
   b) JK master slave flip-flop
   c) The JK flip flop race condition.
   d) Asychronous Counters.
   e) 8 bit ripple counter.
   f) 4 bit synchronous up-down counter.

Q.19 Fill in the blanks:

   a) A circuit that needs only a signle data input is called ....

   b) Synchronous counters are also called as ....

   c) A three flip-flop counter is reffered to as ....counter.

   d) The.. one bit memory device.

   e) The .... counter has a limitation on its highest.

   f) The...is similar to a demultiplexer.

   g) A binary to decimal decoder is also called.................................

h) .... is same as repeatitive addition.

i) Controller Inverter is used to produce ....

   j) ASCII Code is used to...

   k) Parity Bit is used fo...

   l) The expansion of ASCII is...

   m) An...is a semiconductor device which combines
transistors, diodes, resistors and capacitors.

   n) If we invert both the inputs of AND gate, the result will be ....gate.

Q.1   What are the various registers in 8085?

Ans.   The many registers in the 8085 include the Programme Counter, Stack Pointer, Temporary register, Instruction register, and Accumulator register.

Q.2   What are the 16 bit registers in the 8085?

Ans.   Programme counter and the stack pointer both have 16 bits.

Q.3   What different kind of flags are utilised in 8085?

Ans.   Carry flag, Parity flag, Zero flag, Auxiliary flag, Sign flag.

Q.4   What exactly is a stack pointer?

Ans.   The address at the top of the stack is stored in the microprocessor's special purpose 16-bit register known as the "stack pointer."

Q.5   How do I define a programme counter?

Ans.   The address of the next instruction to be fetched for execution, or the location of the subsequent byte of a multi-byte instruction that has not yet been fully fetched, is stored in the programme counter. In both situations, it is automatically increased by one as the instruction bytes are fetched. Additionally, the next instruction's address is stored in the programme register.

Q.6   What stack is utilised in 8085?

Ans.   In 8085, a LIFO (Last In First Out) stack is employed. The most recent information can be fetched first in this kind of stack.

Q.7   When the HLT instruction is carried out by the processor, what happens?

Ans.   The buses are tri-stated as the microprocessor enters halt mode.

Q.8   What is a bus, exactly?

Ans.   A bus is a collection of conductor lines that carry address, control, and data signals.

Q.9   What is Tri-state logic, exactly?

Ans.   High, Low, and High impedance states are employed as the three logic levels. Normal logic levels are high and low, and high impedance states are open circuits in electrical systems. A third line known as the enable line exists in tri-state logic.

Q.10   Give one address microprocessor as an example?

Ans.   A microprocessor with one address is the 8085.

Q.11   How are interrupts categorised in 8085?

Ans.   The interruptions in the 8085 standard are divided into hardware and software interrupts.

Q.12   What are hardware interrupts?

Ans.   INTR, TRAP, RST7.5, RST6.5, RST5.5.

Q.13   What do software interrupts do?

Ans.   RST0, RST1, RST2, RST3, RST4, RST5, RST6, and RST7.

Q.14   Which interruption is the most important?

Ans.   TRAP is given top priority.

Q.15   Identify the five different addressing modes.

Ans.   Instantaneous, Direct, Register, Register Indirect, and Implied addressing modes.

Q.16   How many interruptions are there in 8085,?

Ans.   8085 has 12 interruptions.

Q.17   What is the clock frequency for the 8085?

Ans.   The maximum clock frequency for the 8085 is 3 MHz.

Q.18   What does the RST for the TRAP stand for?

Ans.   TRAP is the name of RST 4.5.

Q.19   Which is referred to as the High Order/Low Order Register in 8085?

Ans.   A flag is referred to as a low order register, and an accumulator is referred to as a high order register.

Q.20   What exactly are input and output devices?

Ans.   Input devices include things like keyboards and floppy discs. Examples of output devices include printers, LED/LCD displays, and CRT monitors.

Q.21   Can an RC circuit serve as the 8085's clock source?

Ans.   If a precise clock frequency is not needed, then the answer is yes. In comparison to LC or Crystal, the component cost is also low.

Q.22   Why is crystal a favoured source of timekeeping?

Ans.   Because of the frequency's excellent stability, high Q (Quality Factor), and the fact that it does not drift with ageing. Most frequently, crystal is employed as a source for timekeeping.

Q.23   Which 8085 interrupt is not level-sensitive?

Ans.   An interrupt that triggers on a rising edge is RST 7.5.

Q.24   What does the term "Quality Factor" mean?

Ans.   An additional definition for the quality factor is Q. Therefore, it is a number that expresses how lossy a circuit is. The losses are lesser the higher the Q.

Q.25   How do level-triggering interrupts work?

Ans.   They are level-triggering interrupts, RST 6.5 & RST 5.5.